Making Woodwork Aids & Devices

Making Woodwork Aids & Devices

REVISED EDITION

Robert Wearing

Sterling Publishing Co., Inc.   New York

Illustrations by Gay Galsworthy

Library of Congress Cataloging in Publication Data

Wearing, Robert.
  Making woodwork aids & devices.

    Rev. ed. of: Woodwork aids and devices. c1981.
    Includes index.
    1. Woodwork—Equipment and supplies.   I. Wearing,
Robert. Woodwork aids and devices.   II. Title.
III. Title: Making woodwork aids and devices.
TT186.W4   1985   684'.08'028   85-12610
ISBN 0-8069-6264-X (pbk.)

Revised Edition
Published in 1985 by
Sterling Publishing Co., Inc.
Two Park Avenue, New York, N.Y. 10016
Distributed in Canada by Oak Tree Press Ltd.
℅ Canadian Manda Group, P.O. Box 920, Station U
Toronto, Ontario, Canada M8Z 5P9

First Published in the U.K. in 1981 as "Woodwork
Aids and Devices."
Published by arrangement with Bell & Hyman, Ltd.
This edition available in the United States, Canada,
and the Philippine Islands only.

Printed in U.S.A.

Contents

Introduction

Aids and devices have probably been invented by woodcraftsmen since the beginning of time. Their purposes were varied. Some eliminated difficult skills, some produced greater accuracy. Others speeded up production or cut costs, generally in time, though sometimes in materials. Another group enabled large numbers of identical components to be made. It is interesting to note that over the centuries similar problems produced similar solutions.

It is inevitable that the reader is already aware of some of the ideas contained in this book. He may even think he has invented some. I hope these are not too numerous and the majority will be of use and assistance.

Fixings A large number of the items have movable components which are required to lock or grip. A number of fixing methods are shown and many are interchangeable so that the reader can make a choice of what is preferable, convenient or feasible to suit his own case. It should, however, be mentioned that though wrong to the purist, a coarse engineering thread tapped in a dense hardwood like beech has a very long life in a tool which is not in frequent use. For frequent use nuts, preferably square, will need to be let in. Nylon knockdown fittings should not be neglected as some of these give a good anchorage for machine, wood or chipboard screws.

The sketches are, in a number of cases, deliberately foreshortened. This is to show the vital details as large as possible. In a large number of cases it has not been thought necessary or particularly useful to give precise sizes. The reader will evolve these bearing in mind his own particular position, that is his exact purpose and the materials and facilities available at the time of making.

Metric/Imperial conversions are generally approximate. In cases where it is important, accurate conversions have been made.

Great accuracy in assembling wooden devices is often difficult and the material is not always as stable as we would like it to be. The slogan, therefore, to bear in mind is 'If you can't make it accurate, make it adjustable'.

HOLDING DEVICES

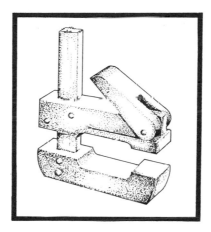

1 Improved sawing board

This board has several advantages over the conventional bench hook, particularly for the beginner and the inexperienced. The board itself is held in the vice thus eliminating one cause of movement. In this position the work can be firmly held with a cramp, further preventing movement. The sawing position is thus similar to the planing position.

Fig.1

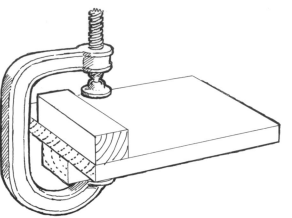

2 Rebated vice jaws

If the basic vice jaws are rebated at their ends, as the illustration shows, quite a variety of useful and helpful auxiliary jaws can be instantly fitted. In a communal workshop it is important that these jaws be accurately and identically machined.

Fig.2

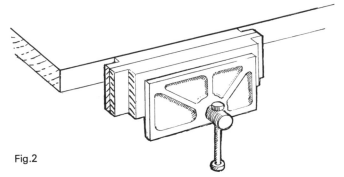

3 Carpet jaws

Good quality 13mm ($\frac{1}{2}$in) plywood makes a good backing for these jaws. It is advisable to machine an ample length of rebated hardwood strip to make the sockets. The sockets are glued on with a rebated jaw in place, using a thickness of card to give an easy sliding fit. Jaws lined with clean carpet offcuts are most useful for holding polished and well-finished work. Great care must be taken to keep these jaws free of glue.

Fig.3

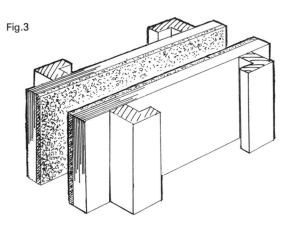

4 Jaws for tapered work

Tapered legs and similar items present a holding problem easily overcome by this single auxiliary jaw. When the taper is cut on a circular saw, the same setting can be used to produce the tapered block for this jaw.

5 Jaws for round work

The two blocks should be cramped or even glued together while the holes are bored. They can then be glued to the plywood in the vice, round material having been inserted in the holes. When marking out for the holes, make sure that they are so placed that when long round material is held vertically, it does not foul the vice slide bars.

6 Cradle jaws

The traditional planing cradle does not hold the work very firmly. The cradle jaws do. They are particularly useful in planing octagons prior to turning or using with rounders, also for planing flats on round material, for example to receive mortises.

Fig.4

Fig.5

make two

Fig.6

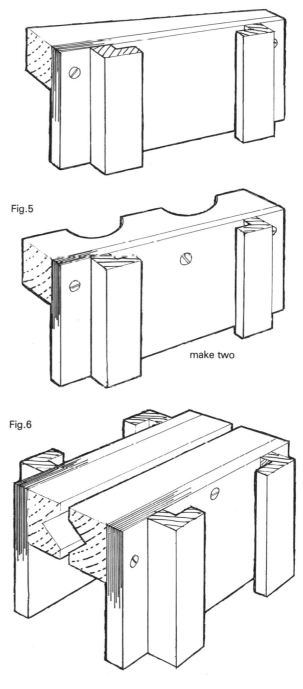

9

7 Leather-lined jaws

These jaws are particularly useful for holding backsaws, small handsaws or scrapers for sharpening. They hold the job firmly and reduce the unpleasant noise which is unavoidable when filing saws.

It is advisable to protect the workbench from metal filings with pieces of paper or a plastic sheet.

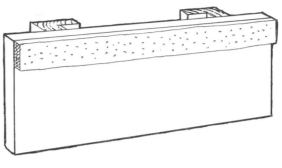

Fig. 7

8 Single tall jaw

This will be found to have several uses. **A** The tall jaw will hold a wide board firmly against a pinned or screwed strip for planing. **B** Squaring wide boards with a knife provides a basic difficulty. The large square tends to drift as the knife runs against it. By pulling firmly on the try square stock both work and square are held firmly against the tall jaw. Squaring at the other end is accomplished by an overhand grip, knifing away from the operator. **C** A block held in the vice is another alternative.

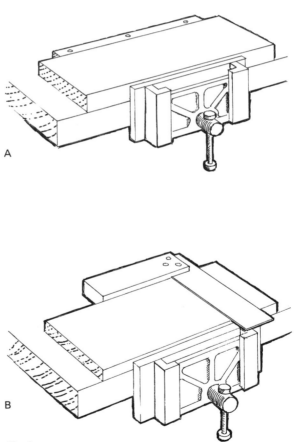

A

B

Fig. 8

9 Sash cramp holder

A convenient way to hold small boards for face planing is to grip them by the ends in a sash cramp which in turn is held in the vice. The plan view **B** shows this. The cramp jaws must, of course, be below the surface to avoid damage to the plane cutter. This jaw arranged similarly to Fig. 3–7 neatly grips the sash cramp bar at one of two heights to suit differing thicknesses of work. The two rebates together must be less than the thickness of the cramp bar.

10 Shooting mitres

Mitres can be shot with this jaw as long as not too great an accuracy is required. The back board is similar to earlier examples and to it is fixed a hardwood strip of about 20 × 13mm ($\frac{3}{4}$ × $\frac{1}{2}$in). The work is gripped firmly against the stop piece while being planed. As long a plane as possible should be used so that a good part of the sole can rest on the bench top.

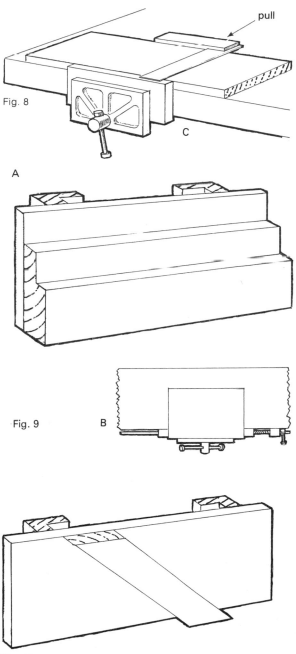

pull

Fig. 8

A

C

Fig. 9

B

Fig. 10

11 Faceplate holder

Bowl turners frequently require to work off the lathe using a sharp scraper on difficult or torn areas. A pair of jaws as illustrated enable one to grip the iron faceplate on which the bowl is mounted. It can be held very firmly while being worked upon and easily rotated for convenience.

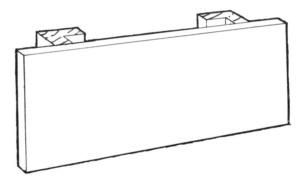

Fig. 11

12 Blank jaws

When machining a length of hardwood for the blocks it is wise to make an extra pair of blank jaws. They are bound to be useful some day for some purpose.

13 Edge planing of long or wide boards

The bench vice holds one end securely, the other end needs support. **A** and **B** show the method when the inner vice jaw is flush with the bench edge. When the vice is not so fitted, for example with the recommended rebated jaws, the device is modified to **C**. Two G cramps hold the device to the bench and the board to the device.

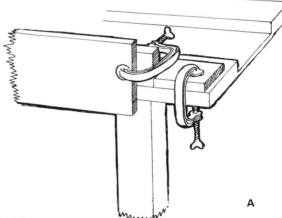

Fig. 12

Fig. 13

A

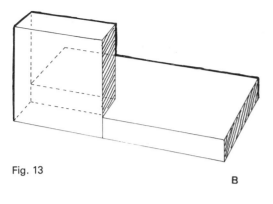

Fig. 13

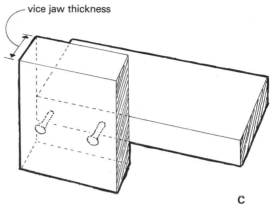

vice jaw thickness

B

C

Fig. 14

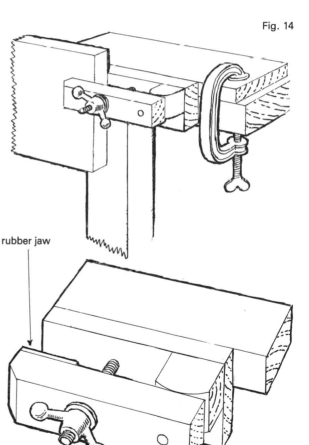

rubber jaw

14 Long board holding device

A further improvement for benches with the recommended outstanding jaws is modelled on the handscrew principle. The baseboard cramps to the bench top with the inner jaw overlapping the edge. This carries a curved spacing block of about 25mm (1in) thickness, into which a short dowel is fixed. A thread of about 10mm ($\frac{3}{8}$in) with wing nut and tommy bar operates the jaw. It is an advantage if the jaw is lined with rubber, for example an offcut of rubber webbing.

15 Mortising block

Fig. 15

It is bad practice to mortise in the vice as the workpiece can be scratched through sliding downwards through the jaws. The vice can be used, however, to grip the mortising block as shown. The work is held to the block with a G cramp or handscrew. This holds the workpiece firmer than cramping direct to the bench top and makes sure it is truly vertical. The thinner the workpiece the greater is the security given by working by this method. Alternatively, the block can be held to the bench top with two long bolts and wing nuts.

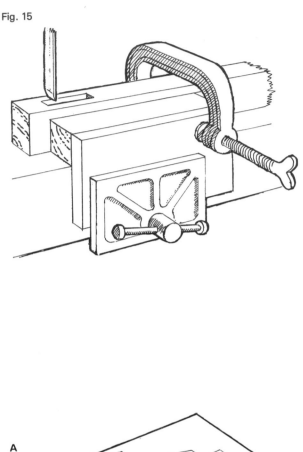

16 Planing boards

This is the simplest of a number of planing boards. Shaped work always presents holding problems, particularly thin pieces. A stout block is gripped in the bench vice. The board itself may be solid wood or blockboard. A small amount overhangs the vice jaw with the bulk sitting on the bench top. Buttons of hardwood, slightly thinner than the workpiece, and fitted with countersunk screws, hold the job firmly, **A**. A further improvement is to replace the buttons with circular cams, **B**. These can be turned to a diameter of, say, 35mm ($1\frac{3}{8}$in) then sawn off and drilled well off centre and countersunk. A 6mm ($\frac{1}{4}$in) hole takes a No. 12 screw.

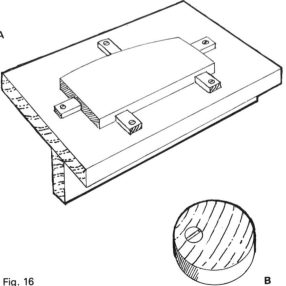

Fig. 16

17 Planing boards

This boomerang type of board will hold both parallel and slightly tapered components for planing and it is particularly useful for holding thinnish pieces. The edge strip and the boomerang must, of course, be thinner than the workpiece. An assortment of lengths, widths and thicknesses is desirable. The boomerang should be of multi-ply and should be well countersunk to avoid damage to the plane iron. The principle is very simple. The more you push, the tighter the grip.

The boomerang can be screwed to the bench top and used with a tall auxiliary jaw as in Fig. 8, or a piece standing above vice jaw height. In this case it is advisable to drill a line of 6mm ($\frac{1}{4}$in) holes about the same depth 6mm ($\frac{1}{4}$in) and about 25mm (1in) apart. This avoids projecting roughness on the bench top caused by the screws.

18 The lever cam clamp

The use of cam blocks on a holding or planing board has already been described, Fig. 16. This variation is frequently found to be better and more convenient. Make them from stout plywood and drill eccentrically, **B** shows a typical shape.

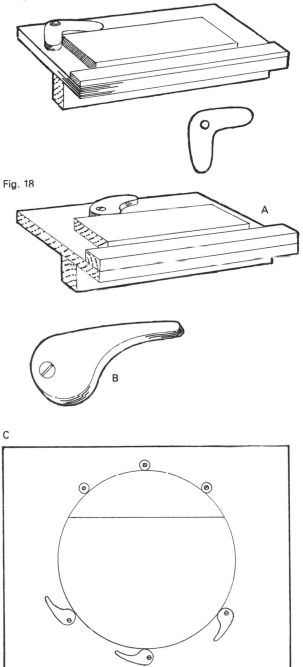

Fig. 17

Fig. 18

A

B

C

15

In **A** a small board is shown being held on a planing board between a lip and a lever cam clamp. Screw the clamp on so as to grip on a smaller radius. Pulling in the lever increases the radius at the grip point. Make sure when fixing that the lever cam clamps even more tightly if the workpiece moves forward.

This device has also been used successfully in gluing up a small circular table top which had split, **C**. Use a 20mm (¾in) blockboard or chipboard base, a few basic circular cams, Fig. 16B, and, say, three lever cam clamps. Providing that the cams are firmly screwed, the lever clamps will bring the joint together and hold it firmly.

19 Clamping sawing board

The sawing board illustrated was designed for a handicapped student having the use of one hand only. The ideal clamp is one of the fast-threaded wing nuts, alternatively a screw and nut must be modified from standard patterns. Backflap hinges secure the clamp plate to the angled rear block. The screw position must permit the gripping of material 75mm (3in) wide. A jaw made from a piece of so called half-round moulding 25mm (1in) nom. wide allows varying thicknesses to be firmly held. A strip of glass-

Fig. 19

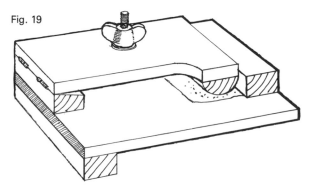

paper on the base further helps with the grip. The right-hand side of the clamp piece is cut away to give clearance to the finger holding the saw handle.

This sawing board has proved very popular with students who are in no way handicapped.

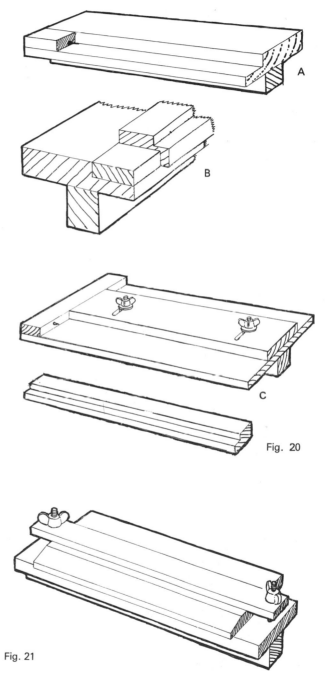

Fig. 20

20 Sticking board

The sticking board, **A** and **B**, is the traditional way of holding narrow components for rebating, ploughing and planing or scratching mouldings. The board can be built up or rebated. A number of such boards will be needed in different sizes. **C** shows an adjustable sticking board which will meet the great majority of requirements. A small clipped off nail helps to hold the work firm on either board but on the finest work this may well have to be removed. The fence may be built up or the slots can be cut out of the solid.

Fig. 20

21 Panel fielding aid

The fielding of panels can be much simplified with this aid. Panels do not always cramp easily to the bench top. This model, if not made too small, will cover a range of panel sizes. It will not only hold the workpiece firmly but will also provide a fence against which a jack or rebate plane can run.

Fig. 21

17

22 A holding method

This method was found convenient for holding chair seats for hollowing and finishing and has since proved to have a number of other applications. A stout block permits two G clamps to grip near the sides. A notched support leg forced into place holds the whole arrangement firm. The workpiece can be rotated to work in any direction.

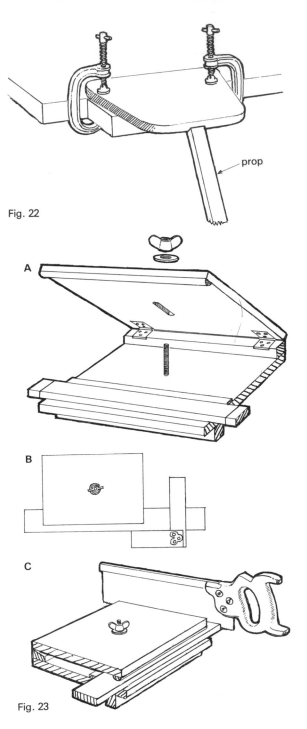

Fig. 22

23 Holding device to square and saw shoulders

Small mortise and tenon jobs can be speeded up and rendered more convenient by the use of this aid. **B** shows squaring a shoulder preparatory to knifing the shoulder line and **C** shows how convenient the sawing is. Both left and right shoulders can often be sawn at the same cramping. **A** is virtually self-explanatory. Solid wood or blockboard may be used and the hinging strip, if 25mm (1in) thick will cope with most work. The stop bar should be thinner than any of the work anticipated. A half-round strip on the hinged member provides a firm pressure over a range of sizes. A 10mm (⅜in) screw and wing nut will give adequate cramping.

Fig. 23

18

24 Holding slightly-shaped work in the vice

Fig. 24

When planing slightly-shaped work in the vice it is inevitable that it will see-saw. This can be prevented by supporting it on a packing piece, or pieces, of suitable thickness which rest on the vice bars.

25 Cleaning up a small table

All that is required here is a stout, nicely-planed board. The table is threaded round the vice and is held in place by this board. If the legs are tapered insert a small wedge under each. In this position the horns can be sawn off and the surface conveniently planed and sanded.

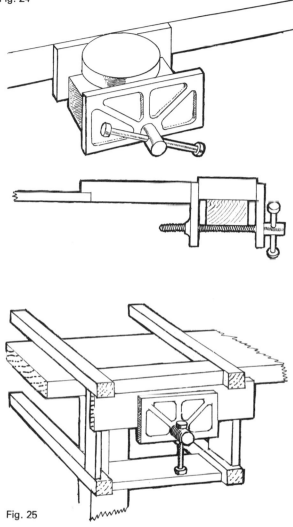

Fig. 25

26 Cleaning up a table top

Any large board can conveniently be held in this manner. Probably three-quarters of it can be comfortably worked from the far side of the bench. Prepare a stout block, slightly angled, and face one side with carpet. The work is cramped to this with two G cramps. When the block is gripped by the vice it will rotate slightly, forcing down the board very firmly at the uncramped end where work will take place.

27 Bridge cramping

An awkward-shaped workpiece on to which G cramps will not conveniently fit can often be held firm by the bridging techique. A block is produced of the same thickness as the job. A stout piece bridges the gap between the two. This can be held with a G cramp or by a strong bolt through a hole in the bench top.

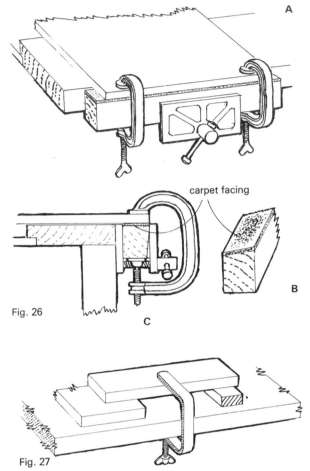

carpet facing

A

B

C

Fig. 26

Fig. 27

28 Picture frame cramps

Many commercial picture-frame cramps deal with only one corner at a time, thus requiring nails. A quality frame with a glued joint reinforced by veneer 'feathers' requires all the joints to be glued at the same time.

The simple type, **A**, can be produced as a rebated strip then sawn off to make four blocks. Chisel or file the notches.

The thread through type, **B**, is easier to use but is more complex to make. Saw and chisel out the small platform, then drill for the cord. A refinement is to drill a hole of about 5mm ($\frac{3}{16}$in) in the corner before sawing. This gives a little clearance right in the corner when gluing up. The platform should be kept well waxed in use.

Both types are tensioned with a double length of thin nylon cord. The four twisting sticks must always be rotated in the same way, for example clockwise. When cramped up check for equal diagonals to ensure frame is square. If not, two V-blocks of thin ply screwed to a flat board will correct this, **C**. Over correct, allowing for spring back.

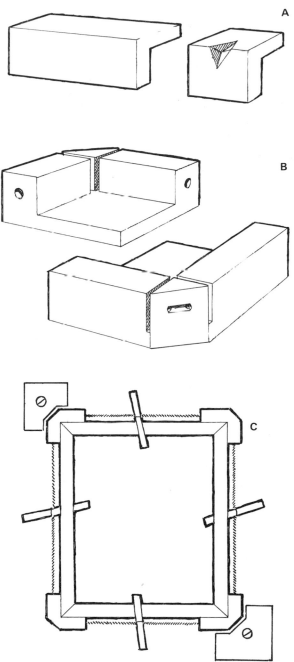

A

B

C

Fig. 28

29 Light picture framing cramps for small frames

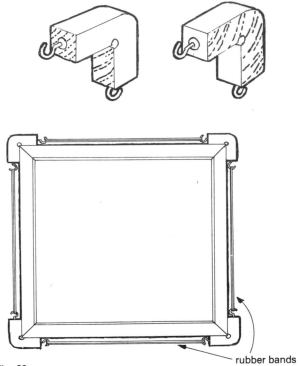

Tensioning by twisted string is not practicable on very small frames. As very little pressure is required, however, rubber bands can be substituted. These corner blocks generally need to be quite thin and the grain direction can be either of those shown. Drill the corners out before sawing. This gives clearance for the sharp corner of the mitre, making sure that the pressure is where it should be and not just on the point. Fit screw hooks and assemble round the frame adding rubber bands until sufficient pressure is built up. Glue up on a sheet of thin polythene on a flat surface such as chipboard.

rubber bands

Fig. 29

30 Screw adjusted light cramps

The illustration shows this further alternative. The screw and wing nut should be about 4mm ($\frac{3}{16}$in). The three plain blocks should have their corners well rounded to permit the cord to move easily. A thin woven nylon cord is suitable. When tightening the wing nut it may be necessary just to grip the screwed rod with pliers to prevent rotation.

31 Mitre cramp and corrector

Mitres in small frames can be accurately made without a shooting board by the use of this aid. It consists of a blockboard or plywood base, under which is fixed a stout block for gripping in the bench vice. Two hardwood blocks with mitred corners are glued and screwed to the top, making an accurate right-angle. The top is protected by a piece of hardboard.

The joint, which has been cut on an ordinary mitre box, can be glued in a straightforward manner using two G cramps. If, however, the joint is not a good fit when cramped up dry, open the joint very slightly, then cramp again. Run a fine saw through the joint, blow out the sawdust and recramp.

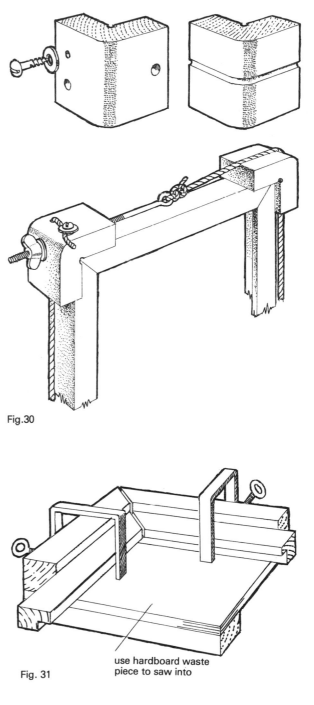

Fig.30

Fig. 31

use hardboard waste piece to saw into

32 Jig to cut keyed mitres

The mitred joints of picture frames, small trays and shallow boxes are very fragile unless strengthened. Strengthening by small nails or pins is crude. A better method is a number of sawcuts into which are inserted small triangles of either matching or contrasting veneer, known as feathers, **A**. The fragile nature of the joint makes it necessary for it to be firmly held during the sawing. This jig, **B**, does just that.

The back plate of multi-ply supports a triangular block with an apex of exactly 90°, **C**. This holds the clamping screw with its wing nut. Two dowels, fitting in loose holes, support the clamping bar and prevent it from rotating. The large base block permits the whole device to be held in the vice yet leaving room for the moulding to thread through. A pair of simple coil springs open the clamp bar to facilitate operation.

Very small frames may not fit on this vice-held device. For them a smaller device has been drawn, **D**. This will cramp or screw to the bench top then operate in the same way.

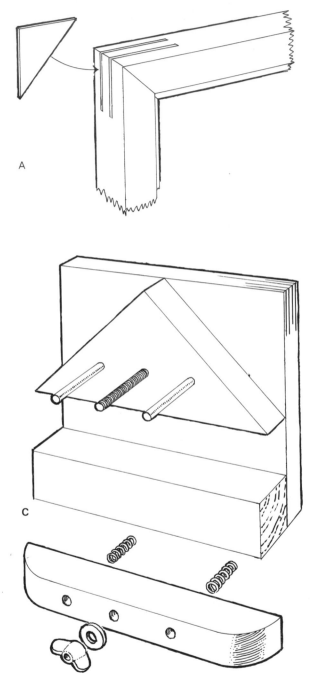

A

C

33 Saddle to hold boxes

Boxes with mitred or lapped joints can have their corners strengthened by means of veneer keys. This technique is particularly suited for boxes which are to be veneered, **A**. Larger boxes can be threaded over the front jaws of the vice. Smaller boxes present a holding problem. The mitred joints are very fragile during the sawing process.

This saddle type jig, **B**, can either be held in the vice or screwed to the bench top. The box can be firmly held with two small G cramps for the sawing.

B

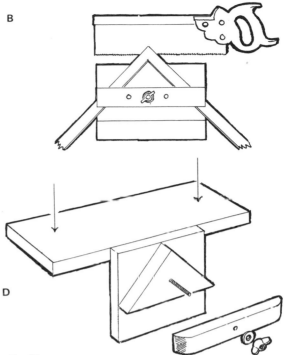

D

Fig. 32

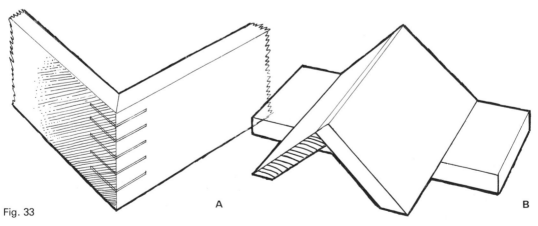

Fig. 33

A

B

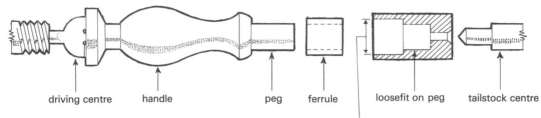

driving centre   handle   peg   ferrule   loosefit on peg   tailstock centre

Fig. 34

loosefit on ferrule

34 Pusher for ferrules

Ferrules, to be successful, must be a really tight fit on the turned handle. The ferrule is made first, from thin wall tubing, then the stub must be turned to a very tight fit. It now requires a pusher to put on the ferrule easily and square.

The pusher is turned from mild steel to the section shown. No sizes are important apart from the diameter and depth of the large ferrule hole. A small through hole is well countersunk to take the tail-stock centre. When pushing the ferrule on to the handle, make sure that the parting off point is not too thin. A sash cramp may be used instead of a lathe for pressure.

35 Box holder

A box, before its top or bottom is fitted, can be conveniently held by the device, **B**, in the vice as in **A**. The central member, approximately the same thickness as the box side, is planed to a very slight taper to give opening jaws.

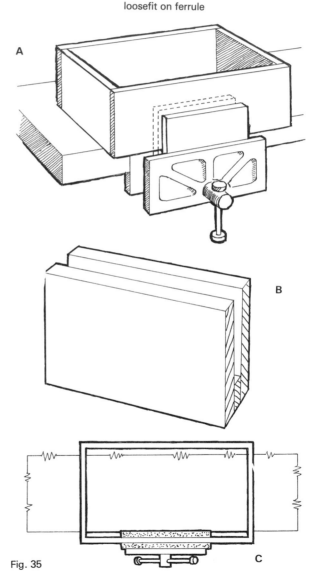

Fig. 35

26

This gives firm holding for working on the top edge, unimpeded by cramps, **C**. Where, however, a bottom has been fitted, a stout board held in the vice, and two G cramps is the best that can be done, **D**.

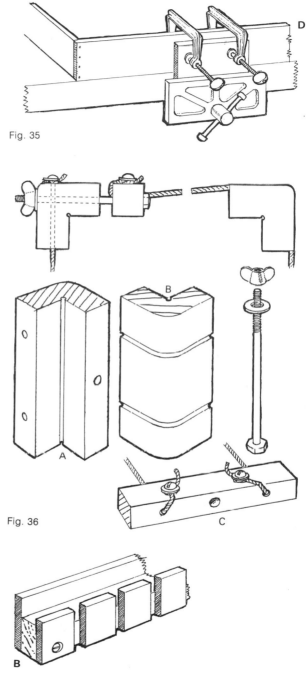

Fig. 35

36 Cramp for small mitred boxes

This is only a development of the picture frame cramp, Fig. 30. Four corner blocks are made to suit the sizes of box commonly made. Three are as **B**, one is as **A** and block **C** is the adjustable anchorage for the cord or wire. The cord holes in **C** match up with those in **A** and the grooves in **B**. Tension is achieved using a 4mm ($\frac{3}{16}$in) screwed rod and wing nut through the central hole in **A** and **C**.

37 Hanging rack for sash cramps

This method holds the cramps firmly and safely, **A**. It allows them to be stored very close together thus economising in wall space which is always short. The block can be cut from one piece or built up and, if built up, the front can be made from 13mm ($\frac{1}{2}$in) ply for extra strength. Illustration **B** makes the idea obvious. The slots should accept the cramp bar easily, the groove behind the front face should accept the foot of the screw end

Fig. 36

Fig. 37

of the cramp. The block holds the cramp very firmly by the feet and very safely. It is virtually impossible for the cramp to fall out.

The lower block, **C**, serves two purposes. Firstly the cramp bar is prevented from damaging the wall. Secondly it serves as a shelf for storing the other blocks described in this section.

38 Sash cramp accessories

Wide boards can conveniently be held for planing by a sash cramp gripped in the vice, where the inner wood jaw of the vice projects beyond the bench edge. Where the inner vice jaw is flush with the bench edge or where it is required to hold a piece smaller than the vice jaws an alternative method is necessary. Several grooved blocks of different lengths are needed. There is no point in making these longer than the vice jaws but one or two smaller ones are often useful. They can be either grooved, **A**, or built up, **B**. In either case the groove should be an easy fit on the cramp bar. One side of the block should be very slightly angled, **C**, so that the cramp bar is firmly gripped when the vice is tightened. The total width must exceed the size of the sash cramp foot. Height will depend on individual circumstances. The arrangement seen from above is shown at **E**.

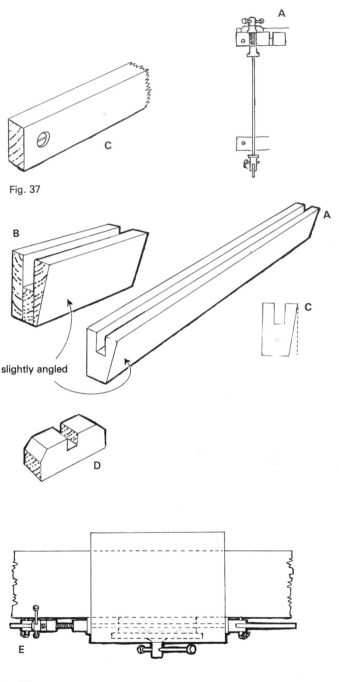

Fig. 37

slightly angled

Fig. 38

Sash cramps invariably topple over at the most inconvenient moment. A set of auxiliary wooden feet, preferably identical, is useful here, **D**. Use a hard hardwood, suitable sizes are about $100 \times 40 \times 22$mm ($4 \times 1\frac{1}{2} \times \frac{7}{8}$in).

39 Aids to large door fitting

The edges of large doors are usually planed by standing astride the door and gripping with the knees, a thoroughly inconvenient position. The method shown in **A** is a considerable improvement. The cost is negligible and the device is easily transported. It consists of one or, preferably, two stout blocks of about 100×75mm (4×3in). A central slot is cut wide enough for the thickest door likely to be encountered. A tapered socket is cut to a depth of about 25mm (1in). A wedge is made to suit this. Well-fitting wedges, tightly driven home, produce a very rigid structure on which it is very easy and comfortable to plane.

The rocker shaped foot lift, **B**, is a great help to anyone hinging a heavy door without assistance. A suitable material is 50×50mm (2×2in) hardwood. Quite thin ends are required with a smooth curve to the centre. The method of use is obvious, as shown in **C**. Other applications will be found.

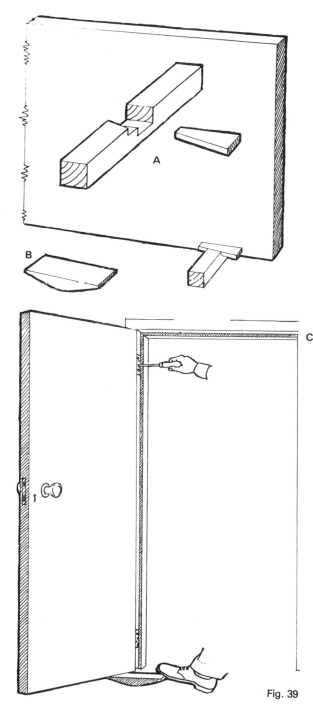

Fig. 39

40 Right-angled holding brackets

The better form of bracket is made from two hardwood strips dovetailed and glued together. A plywood fillet preserves the right-angle exactly, **A**. A simpler and quicker, though not such a good version, is also shown, **B**, made from one solid piece of hardwood with holes bored and enlarged to the shape illustrated to take two G cramps.

Illustration **C** shows a dovetail joint being firmly held for marking. On wide carcase joints a pair of brackets is an improvement. The brackets are also useful for assembling simpler, nailed and glued butt joints and lapped joints and for holding dowelled joints during the drilling.

41 Gluing-up very thin boards

Thin boards, when cramped together with a number of sash cramps, invariably buckle. This method, commonly used by guitar makers is both simple and successful. Having planed the joint, usually on a shooting board, it is laid on a piece of blockboard, with a thin lath 20 × 6mm ($\frac{3}{4}$ × $\frac{1}{4}$in) underneath the joint. Two thin hardwood strips are pressed firmly against the edges and are screwed there as in **A**; the lath is then removed. **B**

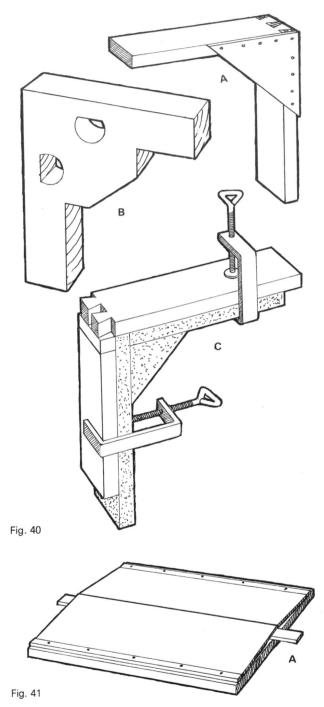

Fig. 40

Fig. 41

30

shows the actual glue up. Glue the joint then return to the block-board with a strip of waxed paper below and above the joint. By means of a stout batten slightly curved in length and two cramps the two boards are sprung to-gether. This method makes sure of a good cramp up and a flat board.

When uncramping after twenty-four hours, first release one of the screwed strips. This will pre-vent the jointed board from springing up.

42 Violin cramps

A large number of these light-weight cramps are needed when gluing on the back and front of stringed instruments. They are cheap and easy to make and have numerous applications in the workshop. The diagrams are self-explanatory. Some round stock is prepared, by turning or with a rounder, to 38mm (1⅜in) diameter. On the sawbench cut up the blocks, then suitably drill for the 5mm (¼in) threaded rod. Some of the blocks need the tapping drill, the remainder are given a clearance hole. Secure the rod in the threaded hole with an epoxy resin glue. Glued on leather pads prevent both damage to the work and rotation.

The lighter model, **A**, adjusts with a wing nut and the handled ver-sion, **B**, is an improvement.

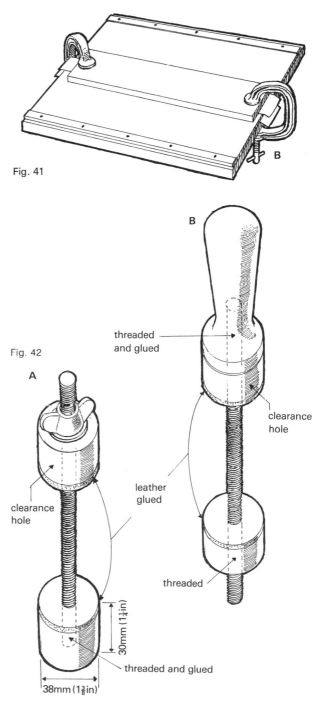

Fig. 41

Fig. 42

43 Guitar maker's cramp

This cam cramp appears to have been developed by guitar makers to satisfy their special needs. They had to be lightweight, have a long reach and, as quite a number were required, cheap. Beech or any close-grained hardwood will make the jaws. Cramp the two together and bore out the ends of the cut out. In sawing the tenon make the spring cut first then wedge this open to get the best possible sawcut.

The bar is from light aluminium alloy. In the moving bar drill for the pins very accurately. To avoid work on the joints, it is in fact possible to build up the jaws from three layers.

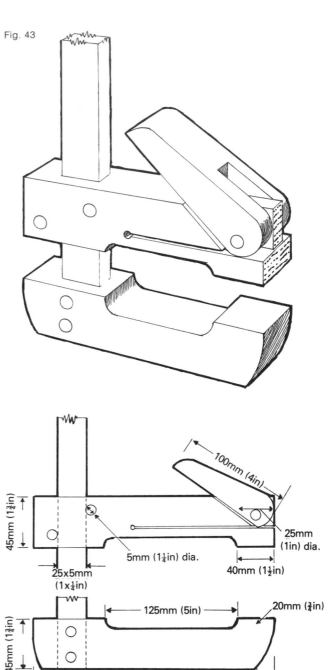

Fig. 43

100mm (4in)

45mm (1¾in)

25mm (1in) dia.

5mm (1¼in) dia.

25x5mm (1x¼in)

40mm (1½in)

45mm (1¾in)

125mm (5in)

20mm (¾in)

225mm (9in)

MARKING
AIDS

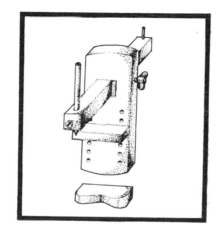

44 Straight-edge

A straight-edge, longer than the normal steel versions of 500mm–1m (2–3ft), is occasionally required. Select the wood carefully using only well-seasoned stable hardwood. Produce to size, then true up the true edge ideally by machine planing. The other edge should be slightly shaped to avoid confusion. Drill a hanging hole and always hang when not in use, leaning against a wall over a period will bow the straight-edge. Varnish or paint well as this not only helps to reduce movement, but by making it readily identifiable as a tool also prevents misuse in a communal workshop.

45 Diagonal laths

A pair of laths, **A**, will check the equality of diagonals inside a frame or carcase. Push firmly into two corners and make the pencil mark shown. Repeat in the other corners and compare. Using this method and measuring the combined sticks enables a length to be measured in a place where a rule cannot be held, e.g. from the bottom of one small hole to the bottom of another.

Model **B** is more commonly used where cramps and other restrictions are not in the way. Make the two pencil marks and adjust the cramps to give the mean reading.

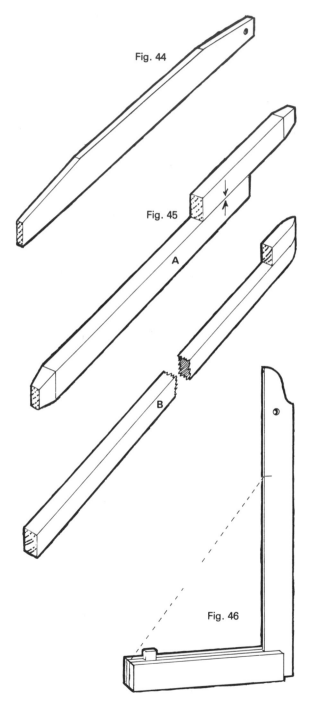

Fig. 44

Fig. 45

A

B

Fig. 46

46 Large wooden square

This is used not so much for checking right-angles, the diagonal lath is more accurate, but for marking out large components of plywood and other sheet material. It should not be used with the marking knife which damages it.

47 Aid for copying angles in buildings

When building in fitments to a room one is soon made aware of the fact that few corners are really right-angles. In order to saw large pieces of plywood, chipboard and similar materials to fit the corner exactly a large copying square is needed and 750 × 600mm (30 × 24in) is a useful size.

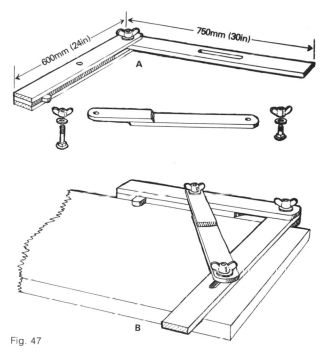

Fig. 47

This can best be made by laminating strips of hardwood or plywood as shown in **A**. The stock is of three layers, the centre one stopped well short of the rounded end. The blade is of a single layer also with one rounded end. Cut a short 6mm ($\frac{1}{4}$in) slot in the centre of the blade and drill a 6mm ($\frac{1}{4}$in) hole in the centre of the stock and a same-sized hole in the curved ends of both. A 6mm ($\frac{1}{4}$in) bolt with wing nut makes the pivot. A diagonal brace is necessary in order not to lose the angle during handling and this is made of the same material again with the ends drilled and rounded. A small stud inserted into the centre layer prevents the square from tilting in use.

Push the device hard into the corner to be measured and preserve the angle by fastening the brace. This must be fixed to the side convenient for the subsequent marking on the board. Illustration **B** shows how the angle is transferred to the board.

48 Try-square pencil gauge

This is a useful tool for rough marking out in the early stages. Many try squares are already calibrated, making the work that much easier. Drill holes at convenient intervals, 10mm or $\frac{1}{2}$in, of a size to accept either a pencil point or a ballpoint pen.

Fig. 48

49 Glue up square

This useful tool for checking angles when gluing up is simply made from truly flat plywood grooved into a wood stock. It is very robust and should stay accurate. Be sure to wash off any glue it picks up immediately.

Fig. 49

50 Setting out square and gauge

This is useful for setting out full-size designs or full-size details of larger work. The stock can be chopped from the solid or, more accurately, built up. It should be a tight working fit on a metre (3ft) rigid rule. A hole of about 25mm (1in) is bored in the top side and a well-fitting plug is turned to fit it. One surface is faced with fine glasspaper. With the rule in place the plug is dropped in, glasspaper down, and a square of thin rubber is glued over the top. This tool can be used either as a T square or, locked by pressure on the rubber, as a pencil gauge.

Fig. 50

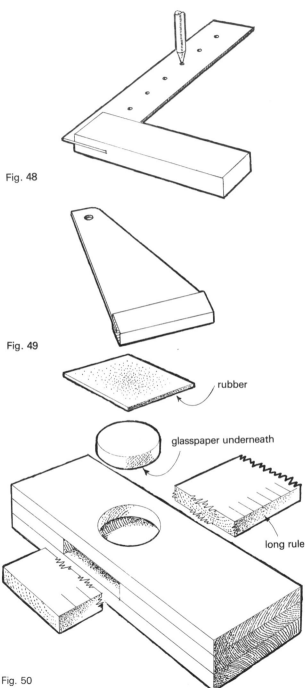

rubber

glasspaper underneath

long rule

36

51 Four-way block gauge

Made from small pieces of very hard hardwood several of these small block gauges prove very handy. Four rebates can be cut each increasing in size by 1mm ($\frac{1}{16}$in). They are used with pencil or ballpoint pen for rough marking out or in situations where the conventional marking gauge is unsuitable.

52 The double chamfer gauge

This is similar in style to the block gauge. Rebates are cut with an allowance for pencil thickness. The corners are rounded to a quarter circle to mark stopped chamfers.

53 Depth gauge

This can be made in various sizes and fitted with a dowel, pencil or ballpoint pen. Alternative clampings are shown in **A**, **B** and **C**. The larger sizes are handy to gauge the depth when turning bowls. Small versions fitted with a ballpoint pen or pencil are useful with a router. In making housings, as at **D** overleaf, the cut is continually increased until the gauge will no longer mark.

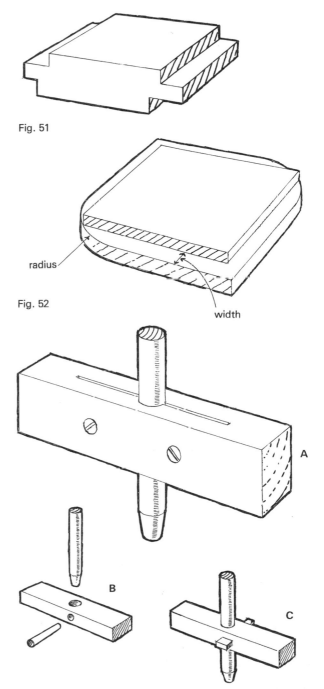

Fig. 51

Fig. 52

radius

width

A

B

C

Fig. 53

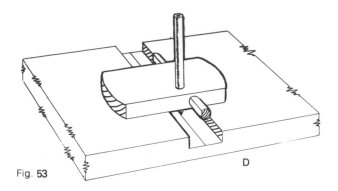

Fig. 53

D

54, 55, 56 Special gauges

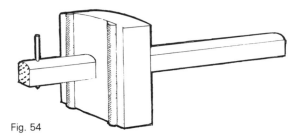

Fig. 54

54. The common marking gauge can be modified by the addition of two half-round guide strips. These permit the gauge to operate successfully on curved edges.

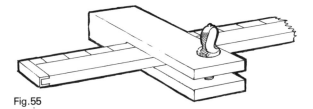

Fig.55

55. A rule gauge for pencil work. The sliding block can either be cut out or built up. This device is particularly useful for marking out wide sheet material.

56. A small steel cutter shaped to fit the end of a normal marking gauge and screwed into place permits marking in otherwise inaccessible places such as corners.

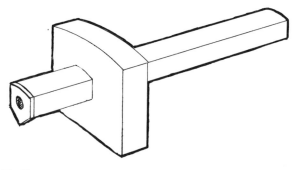

Fig.56

38

57 Panel gauge

It is customary for panel gauges to run on a rebate as it is not the practice to roll the gauge over, as is the method with the marking gauge. Three possible stem ends are shown. When using a gauge point, **A**, the end is thickened with a section of wood to match up with the rebate. If a pencil is used this is not necessary. The pencil may be held with a machine screw, **B**, or a wooden wedge, **C**. In the latter case it is convenient to slot the end on the circular saw and fill in to make the sloping mortise. The stem can be locked either with a wedge, **D**, or a screw, **E**. If the screw is chosen it must grip on a brass shoe as shown.

58 Pencil gauge for curved work

This is similar to Fig. 57 except that the rebate is replaced by two dowels. Suitable sizes are approximate. Stock – 105 × 75 × 21mm (4 × 3 × $\frac{7}{8}$in). Stem – 18–20mm ($\frac{5}{8}$–$\frac{3}{4}$in) square.

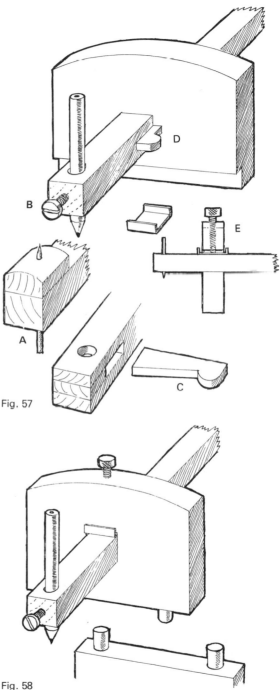

Fig. 57

Fig. 58

59 Combined 'grasshopper' and deep gauge

The complete gauge, **A**, is assembled for what is often referred to as 'grasshopper gauging'. Normally a pencil or ballpoint pen is fixed in the stem although a point end can be used. The auxiliary fence is secured with round-headed screws through a suitable pair of holes. **B** shows the gauge in use. A box with an, overhanging top tacked on is being gauged to give a line on which nails or screws are to be driven. A second, curved, auxiliary fence enables the gauge to work round a curve. Without either fence the gauge becomes a deep gauge. Using a pencil or ballpoint, gauging can be carried out over a step or lipping, **C**, or down into a cavity.

60 Gauge with additional pencil end

The stem of an ordinary marking gauge has been reversed, **A**. The other end, as in **B**, has been bored, filed up as shown and fitted with a small machine screw.

61 Gauging set-in shelves

Frequently it is required to gauge the mortises and tenons on shelves set in behind the front

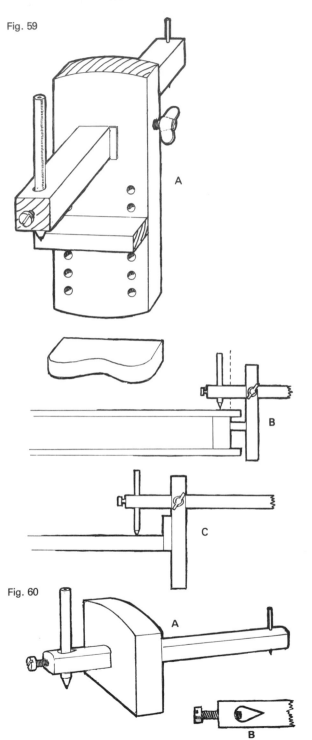

Fig. 59

Fig. 60

edge of the carcase side. A typical jointing arrangement is shown with the customary housing omitted for simplicity.

Decide the amount of set in and prepare a block that thickness as at B. Any accurately thicknessed hardwood will do but multiply is excellent. In use the block is slid onto the stem and the combined gauge is set for the shelf tenon. Afterwards the block is removed and, without adjustment the mortises in the carcase side marked. If well made these blocks are worth keeping to be used for the setting of rails for table or stool construction, etc.

62 Gauging tapers

This is a convenient aid for gauging such items as legs tapered on the inside. It consists of a long hardwood member about 20mm ($\frac{3}{4}$in) thick on to which is dovetailed a short member of the same material, **A**. If the long member is made wider, as indicated by the dotted lines, the whole can be easily held in the bench vice for gauging. A small packing block is made to suit the taper required and the whole is held by hand or lightly cramped for gauging. The right taper can be accurately repeated any number of times. A particularly useful application is the gauging with a mortise gauge of angled tenons such as are used in chair making shown in plan at **B**.

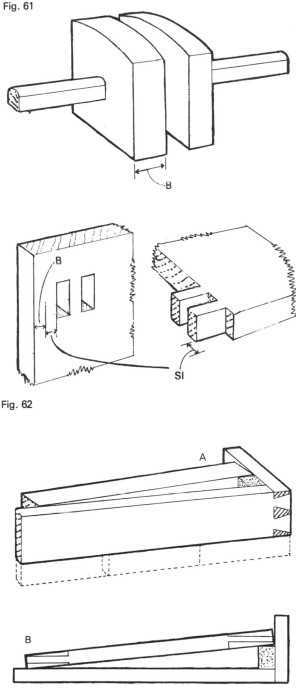

Fig. 61

Fig. 62

63 The pitch stick

The back legs of a set of chairs will require the same amount of rake. While the first leg can be accurately set out from a square-edged board the remaining legs nest inside each other and are drawn from a plywood pattern. Having planed these as accurately as possible, the pitch stick is required for planing the central shoulder which takes the seat rail very accurately. It consists of nothing more than a length of seasoned, accurately-planed hardwood and a dowel in a tight hole. The dowel is tapped out to give the pitch required and the shoulder carefully planed to give the result as illustrated.

Fig. 63

64 The marking out of cylinders

Table, chair and stool legs are typical cases where it is necessary to mark identically the distances between the holes to be bored for rails and also to make sure that the centres of these holes are in a straight line. Two straight-edged strips are glued and pinned together with an end stop forming a type of cradle, **A** and **B**. The required distances are marked on the top edge of the cradle measured from the end stop. Ruling along the top edge easily produces centres accurately in line, **C**. Strong rubber bands are adequate to hold the component still while marking.

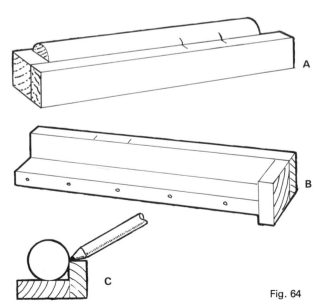

Fig. 64

65 Curved components

Curved components present a holding problem when marking out using the jig described in Fig. 64. The easiest solution is to make an identical pair of wooden collars which will fit on each end of the component and in this form it can be conveniently held in the jig.

Fig. 65

66 Gauging cylinders

Cylinders require gauging in length to mark a straight centre line for drilling or for marking in-line mortises. Chair, table and stool legs are cases in point.

First a cradle is made and this is most easily constructed from two sections glued together with a base block which permits the cradle to be held in the vice. The work can be firmly held by a suitably placed G cramp, gripping on a shaped block as indicated by the dotted line in the diagram. A special gauge is required with an unusually deep stock. A tapered dowel wedge is quite adequate to lock the stem. Suitable stems can be arranged to take one point, two points for mortise gauging, or a ballpoint pen. In use make sure that there is good contact between the stock and the cradle.

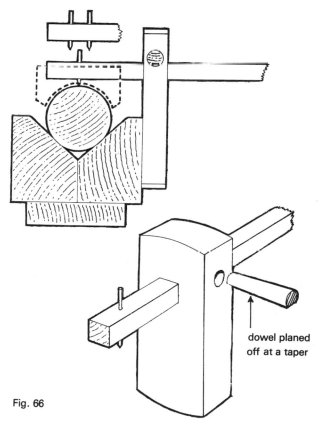

dowel planed off at a taper

Fig. 66

43

67 Matching curves

Lines can easily by drawn to copy curves, **A**, by turning up several suitable discs from thin plywood, acrylic or plastic laminate. In the lathe drill a central hole to take a fine ballpoint pen, **B**. The two methods of use are shown at **C**. On the right the disc works outside a template. On the left it works inside either a template or a bent lath. **P** is the pen position.

68 Marking curves

Curves are best marked along a curved lath lined up over three marked points. There are three suggested methods. A lath of even thickness is bent to the curve required by nipping the ends in a sash cramp, **A**. A well-chosen lath of even thickness, bent by a twisted string in the manner of the bowsaw, **B**. Finally the most rigid and reliable of the three, a small pointed block, **D**, positions the lath at its centre, **C**. The two ends are pulled back to the marks and cramped there.

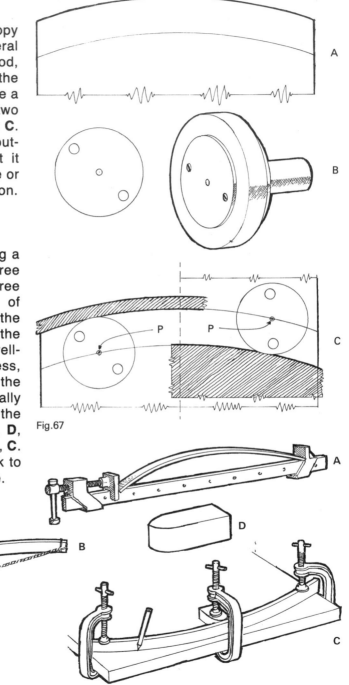

Fig.67

Fig. 68

44

69 Alternative gauge construction

Most of the gauges described already can also be made by this method. It has the advantage that no mortising is required. Using a ballpoint pen means that it is not necessary to roll the gauge on to its corner as with the normal marking gauge. The step, therefore, is quite a convenience.

First prepare stem material of about 20 × 20mm ($\frac{3}{4}$ × $\frac{3}{4}$in), sufficient for a long and a short stem with some extra. Two small offcuts with well-squared ends can be glued between two wider blocks to make the stock. A well waxed offcut can be used as a spacer to give a true square hole. More of the same material can thicken up the stem to carry a normal ballpoint pen. This stem and likewise the stock are drilled and slotted with a fairly wide sawcut.

The clamping screw can be made by threading or soldering a 5mm ($\frac{1}{4}$in) wing nut to a No. 12 brass woodscrew. A normal woodscrew will suit in the stem. Drill the correct clearance and pilot holes. In a good hardwood these threads will last a very long time. A plain stem may be exactly drilled to take a ballpoint refill. Cut the rebate and shape the ends. A shaving may be needed from the stems to give a working fit in the stock.

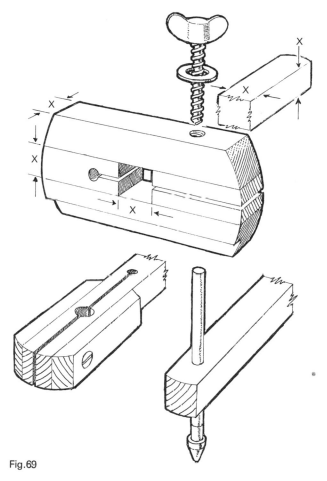

Fig.69

45

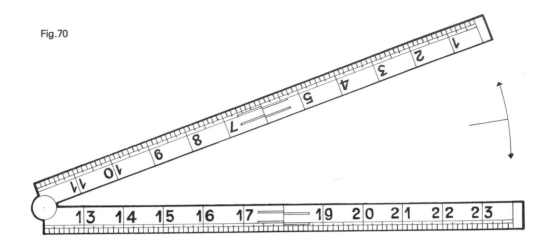

Fig.70

70 Setting out angles

The folding two feet rule enables the setting out of commonly used angles with considerable accuracy. Simply look up the distance in the table and set up the rule as in the table.

Angle	Distance	Angle	Distance
10°	$2\frac{1}{8}$in	40°	$8\frac{3}{16}$in
15°	$3\frac{3}{16}$in	45°	$9\frac{1}{4}$in
20°	$4\frac{3}{16}$in	50°	$10\frac{1}{8}$in
$22\frac{1}{2}$°	$4\frac{11}{16}$in	60°	12 in
25°	$5\frac{1}{16}$in	$67\frac{1}{2}$°	$13\frac{5}{16}$in
30°	$6\frac{1}{4}$in	90°	17in

71 Marking identical angles

When angles other than 90° are to be marked a sliding bevel is usually set. Apart from the difficulty of setting really accurately, a drop or a knock will lose the angle and resetting later in the job is difficult. An altogether more satisfactory method is to make a tapered block to suit the angle and interpose this between the job and the square. In this way the angle can be accurately repeated as often as necessary.

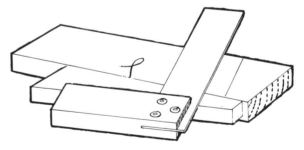

Fig.71

72 Scribers for levelling table legs

However carefully a table or chair is glued up, when placed on a level surface there is invariably some wobble. A scriber can be used to ensure equal lengths and two types are shown. The rectangular one, **A**, has a marking gauge point at each end. By careful positioning of the points at different distances from the four faces eight different scribing heights are possible. The circular model, **B**, takes more making but gives an infinite number of scribing heights.

To use, stand on a truly flat surface such as the circular saw table and level up the table or chair with a wedge under each foot, making sure with a long rule that the top rail is at an equal height at each corner. Now scribe a line on four sides of each leg leaving the smallest amount which can conveniently be sawn off.

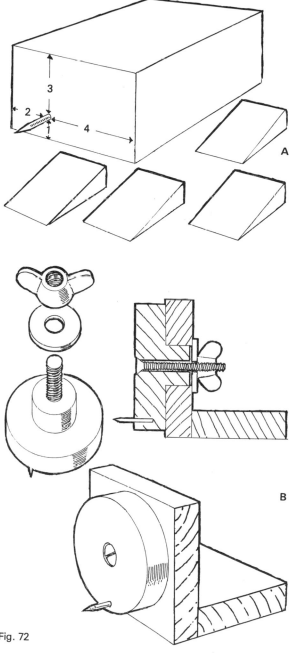

Fig. 72

73 Dovetails – setting the sliding bevel

Dovetail angles for softwoods and hardwoods vary between 1 in 6 and 1 in 8. The most convenient way to set sliding bevels to these angles is by the use of a dovetail board kept in the workshop. Paint a suitable piece of 9mm (⅜in) plywood a light distinctive colour, drill holes for hanging and mark out the slopes with a fine point permanent felt marker, **A**. Note the 1 in 7 bevel right at the end. This permits a bevel to be set at the nearly closed position for marking dovetail halvings.

The angles of dovetail halvings are generally marked with a bevel from the end. This is not ideal since the bearing surface is always small and often not very true. The marking is better done from the edge using the bevel set as described, or better still using the special marker illustrated at **B**. This is made from a short length of hardwood with a slot into which is glued a piece of plastic laminate. When dry, plane or file to the required angle – usually 1 in 7. If, when using this tool, the angle is marked to a short distance from the corner, **C**, the waste can be sawn off thus obviating the less accurate and slower paring from a corner.

The basic dovetail marker can also be made from plastic laminate using an impact glue and pins, **D**. This gives a thinner tool than a wooden one and is easier to make than a metal one.

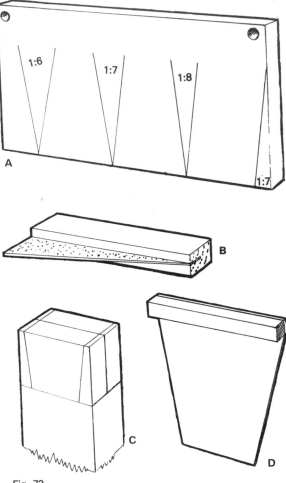

Fig. 73

74 Dovetail marking system

There are many types of dovetail marker and as many methods of using them, but I like to think that this marker, **A**, and its method is pre-eminent.

The plate is made from a piece of 20 gauge brass, steel or alloy 140 × 75mm ($5\frac{1}{2}$ × 3in). This should be sawn and not cut with snips as these may permanently distort it. Drill the four 5mm ($\frac{3}{16}$in) and three 1mm ($\frac{3}{32}$in) corner holes, then sufficient smaller holes to get in a hacksaw blade to cut out the dovetail angle and the two slots. File to shape with great care, rounding the external corners and softening the edges. From an accurate centre line, file the two nicks as shown in **B**. The bar is a piece of brass channel 10 × 10 × 1.5mm ($\frac{3}{8}$ × $\frac{3}{8}$ × $\frac{1}{16}$in) and it has one threaded hole and a slot. The hole is threaded 2BA. A sliding block is fitted below the slot and this, too, is threaded 2BA. The bar is secured to the plate, preferably with a pair of 2BA electrical terminals or with wing nuts and washers, **C**.

Prepare the components and gauge to thickness on the piece which is to contain the tails, **D**. Choose a suitable bevel-edged chisel which is only slightly smaller than the chosen size of pin, **E**. Place the gauge on the work and adjust the bar to give a

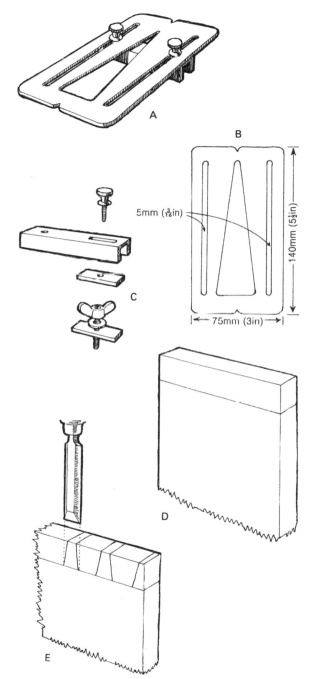

5mm ($\frac{3}{16}$in)

140mm ($5\frac{1}{2}$in)

75mm (3in)

Fig. 74

49

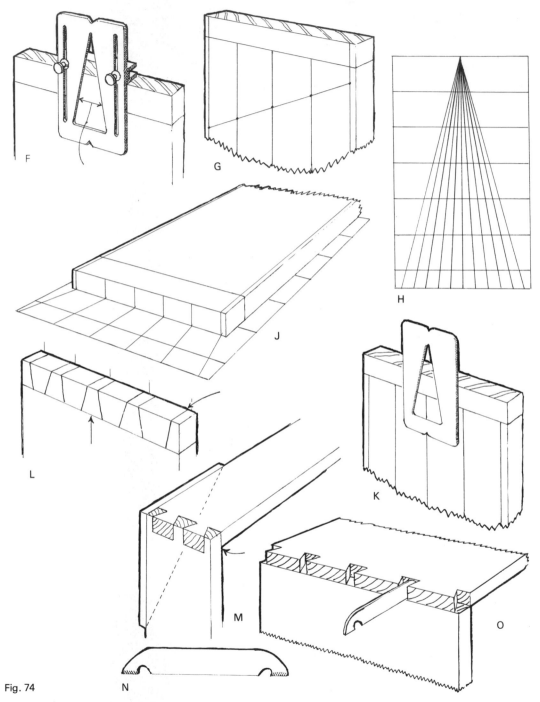

F

G

H

J

L

K

M

N

O

Fig. 74

50

width on the gauge line only slightly more than the chisel size, **F**. True up the bar with the edge of the gauge, either by a try square or against the edge of the wood. For angled dovetails use a sliding bevel on the angle already on the wood.

The two outside or half pins must be a little larger than half size, so, with a pencil, gauge mark down the two outside centre lines to achieve this effect. About 3mm ($\frac{1}{8}$in) is usual for medium-sized dovetails. The distance between these two lines is divided equally in the standard way to give the centre lines of all the pins, **G**. Alternatively a rapid calculator can be made to fix centre lines by drawing equi-spaced converging lines on a piece of plywood as in **H**. The horizontal lines are parallel, and the end of the piece of wood is offered up to the required number of converging lines and kept parallel to the horizontals. The centre lines are then easily picked off, **J**.

The gauge is now placed on the wood with the nick over each centre line in turn for the pins to be marked, ideally with a fine ballpoint pen, **K**. Shellac brushed on gives the surface a 'bite' on which to mark. On the end grain the square marking is done with a marking awl which provides a small groove into which the chisel can be put decisively should there be any error in sawing. Beginners can run a thick pencil into it. Two pencil lines will be formed, and one line should be removed with the saw, **L**. The two pieces are held together for the marking of the second piece, **O**. A pair of holding brackets, **M**, is useful for this process see Fig. 40; one will do for small work. The two pieces are held to the bracket with light G cramps.

A dovetail marking knife is essential for fine dovetails but is useful for all sizes, **N**. It is made from a piece of tool steel 100–125mm (4–5In) long by 12 × 1·5mm ($\frac{1}{2} \times \frac{1}{16}$in) and a piece from a power hacksaw blade is very suitable. Both the little bevels are on the same side, giving a left-hand and a right-hand knife. In use, **O**, the flat side is held hard against the dovetail, and this tool will get into the most inaccessible corner where the marking awl or pencil never got in the past.

The holding bracket, **M**, has a further use when the front or rear corner is to be mitred to take a groove, rebate or moulding. The mitre is often, though not necessarily, at 45° but if the pieces to be joined are not equal thicknesses, the angle cannot be 45°. What is important, however, is that the sum of the two angles should be 90°.

The remaining six diagrams, **P–U**, show how, when the pieces are held at 90°, both angles can be marked from the same edge, using a sliding bevel.

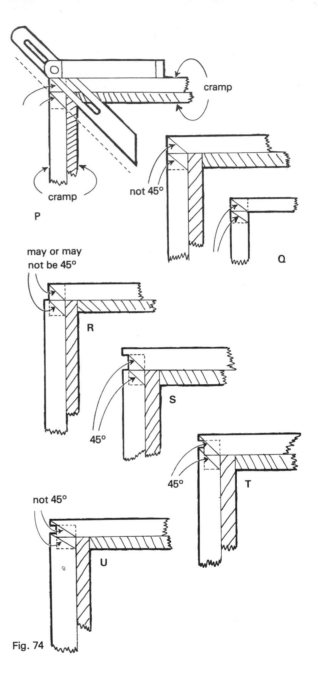

cramp

cramp

P

not 45°

Q

may or may not be 45°

R

S

45°

not 45°

45°

T

U

Fig. 74

75 A dowel joint marker

This is an extremely useful aid for marking out dowelled joints, particularly in flat frames and tables. A piece of plastic laminate is sandwiched in a hardwood stock, **A**. The slot in the hardwood can be cut with a very thin circular saw or a sandwich is built up by gripping the blade in an engineer's vice, holding the wood blocks firmly down on the jaws while gluing. The glue joint is later strengthened by screwing, **B**. Mark one end of the stock black for easy reference.

To use, hold the tool on to the workpiece and draw round the latter. Draw the centre line and mark the dowel positions, **C**. In some cases it will be preferred to stagger the holes on each side of the centre line, **G**. Drill small holes to suit the available marking awl. Note whether the marking began at the black or white end and then line up this chosen end with the edge of the workpiece and prick through, **C**. Reposition on the other half of the joint and repeat, marking out the complete joint as in **E**. The opposite end of a frame is marked out by turning over the marker.

When insetting the rail on a stool or table as in **F**, simply insert a wood block of the thickness of the set-in when marking out for drilling, **D**. The marker can be used many times before it has too many holes drilled in it. After the first use, mark subsequent drillings by ringing with a felt-tipped pen, wiping off when the job is complete.

76 Holder to mark out table legs

Apart from freeing sash cramps this is a convenient device, particularly so in a communal workshop or when numbers of similar tables are being made. Any man-made board will serve for the base with the usual vice strip underneath. A permanent lip is glued on one edge. This must be considerably less than the thickness of the legs, in order that the square can operate conveniently. Two pairs of wedges complete the job, and hold the work firmly while the marking is carried out.

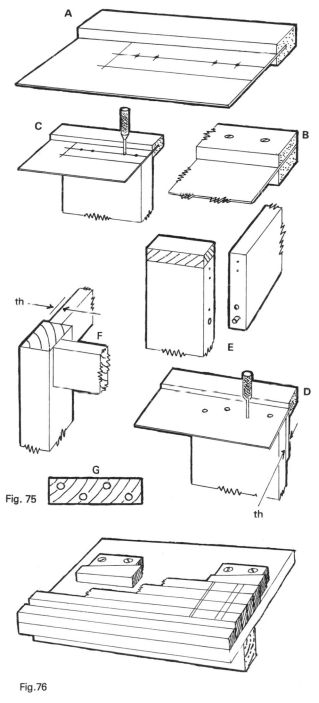

Fig. 75

Fig.76

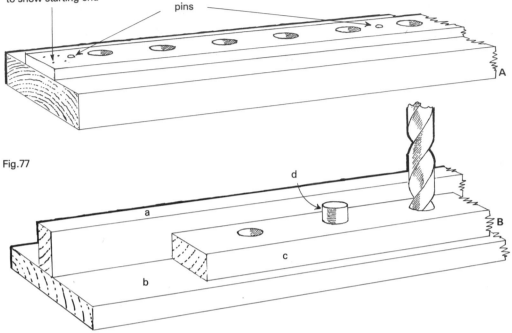

index mark with centre punch
to show starting end

pins

A

Fig.77

d

a

B

b

c

77 Drilling a line of holes

Bookcases and cupboards with adjustable shelves are some of the jobs which require identical lines of equidistant holes.

The illustrations show two methods of achieving this. **A** shows a mild steel strip of, say, 20 × 3mm ($\frac{3}{4}$ × $\frac{1}{8}$in). This is carefully marked out with dividers and drilled. At intervals smaller holes are drilled and through these the guide is pinned to the workpiece. The drilling can now be done by hand drill, by electric drill or by drilling machine. Bear

in mind that after a number of jobs the mild steel will wear and the holes will become sloppy. Do not enter the drill when revolving as this rapidly increases the wear. Identify one end of the guide with a file or punch mark so that the same end is always positioned at the start of the work.

The second method, **B**, is only suitable for the drilling machine. The base **b**, which may be solid wood or blockboard, is planned to take the width of the component in hand. A hardwood fence, **a**, is fixed to the base leaving space behind it for cramps to grip. The workpiece, **c**, is held against the fence and the first hole is bored through into the base. A wood or metal dowel, **d**, is put into these holes. Work and jig are now lined up for the second hole. With the drill still in the hole cramp down to the drilling table.

Drill the second hole then position it on the dowel. This lines up the work for the third hole. This in its turn lines up for the fourth and so on. It may be found convenient to glue in the dowel. This being a tight fit, it need not necessarily protrude through the work. After much use the wooden dowel may wear loose so a metal dowel, or a spare drill of the required size may be thought an improvement.

78 Veneer strip cutter for chess board

The base board of plywood or blockboard has a strip screwed to it for gripping in the vice and two machine screws are fitted at a centre distance of 430mm (17in) as at **E**. A hardboard strip, **D**, is drilled to fit over the machine screws. It is about 75mm (3in) wide. The veneer strip, **C**, has been cut to length to fit in between the screws. The steel cutting guide, **B**, 50mm (2in) wide, forms the top layer. It should first be gripped in a sash cramp which is tightened to bring on a bow which will give a good grip at the centre of the strip. The

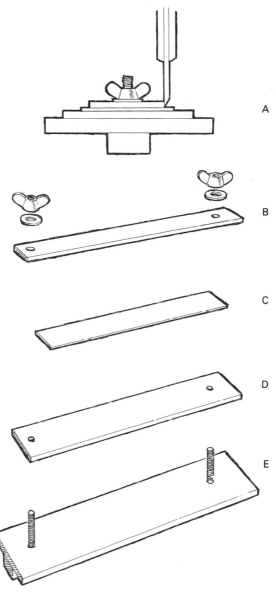

Fig. 78

veneer strips, firmly held under the steel guide can be cut accurately with a sharp knife, bevel on the waste side, **A**.

Four white and four black strips are taped together as in **F**. These are then put through the device again and cut into strips, **G**. Alternate strips are reversed and assembled to form the complete pattern, **H**.

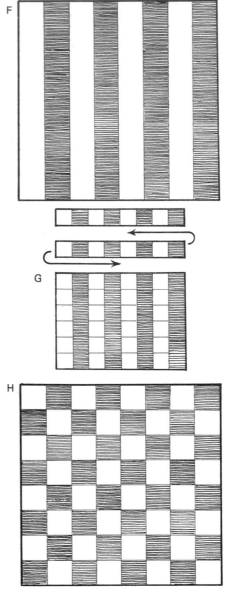

Fig. 78

TOOLS

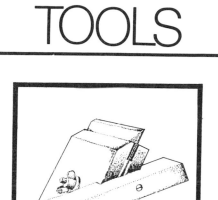

79 Oil pad

'Park your plane on its side lad'. This is a folk custom dating back to the age of wooden planes. The blades of these planes, were firmly held by a tightly hammered in wooden wedge. Following this advice however will disturb the lateral setting of an iron plane whose blade is nothing like so firmly held. Instead park the plane on the oil pad made by gluing a strip of carpet to a plane-sized board. This is a tidy arrangement which both protects the blade and reduces friction. Very little oil is needed.

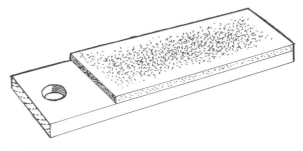

Fig. 79

80 Winding strips

This is probably the most essential of all the tools which cannot be bought and **A** shows one of a variety of patterns. Two identical strips are produced, exactly parallel, from well-seasoned, stable hardwood. They are tapered in section to produce a stable base. The rear strip remains light coloured. Two identical dark blocks are glued on or painted on at the corners. The whole of the front face is painted, stained or veneered to match the corner blocks. So is the top edge.

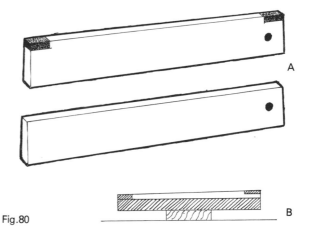

Fig.80

Winding strips are in fact magnifiers of twist or wind. They are placed on what is hoped will be a true face and viewed from a low viewpoint as in **B**. Any twist is shown as a white line at one side between two blacks.

Other design variations can be achieved by the use of inlay stringings. Suggested sizes would be 400–450mm × 50mm × 15mm (16–18in × 2in × $\frac{5}{8}$in).

81 Cabinet scraper sharpening aid

The state of the jaws of the average woodworker's vice make it far from the ideal way of holding the thin scraper when filing and burnishing the edge. This simple device, **A**, holds the scraper firmly and conveniently and in addition eliminates the very unpleasant filing noise. It can be seen from the section, **B**, that there are two hardwood blocks with leather jaws which are kept apart by a thin and very slightly tapered slip glued between them. The height should be such that the device will rest on the vice slide bars with the scraper edge at a convenient height for filing and burnishing.

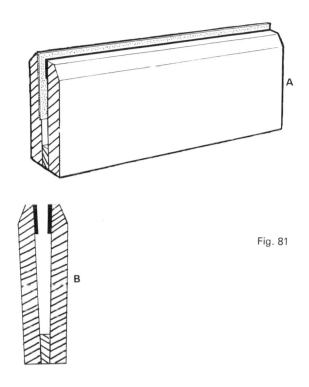

Fig. 81

82 Scraper plane sharpening aid

As these blades need filing not at 90°, as the scraper, but at 45°, the device which was suitable for the scraper has been modified to hold the blade at this angle while still filing horizontally. The jaws are narrower to suit the blade and as they are not closed by the vice, a tightening bolt and wing nut are fitted. A short bar is secured to one jaw (not both) at 45°. With this bar resting on the top of the vice, the blade is held at 45° for horizontal filing.

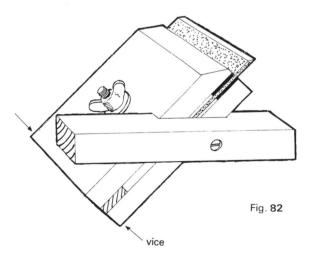

Fig. 82

vice

83 Scratch tools

The type of scratch tool most commonly used is illustrated at **A**. The block is quite straightforward being in two parts, each about 13mm (½in) thick and screwed together. Cutters can be made from pieces of power hacksaw blades, old handsaws or cabinet scrapers. They are filed to the shape of the moulding or groove required at an angle of 90° so as to be able to cut in both directions.

Model **B** is an improvement in that it has a comfortable handle and the fence is easily adjustable rather than the cutter. Model **C** works on the lines of the marking gauge. Its fence, however, is much wider which gives greater accuracy when cutting at some distance from the edge.

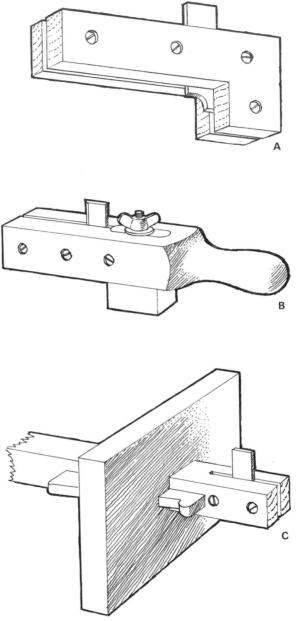

Fig. 83

84 Improved scratch tool

Scratching an ovolo moulding or a small rebate near the edge of a piece presents no problem since, if the tool slips outwards, no damage is caused. When scratching an inlay, for example, the same distance in from the edge, any outward slip will damage the work. The second fence will prevent this on all components within the capacity of the tool.

Mark out and cut the mortises before sawing off the fences and shaping. Prepare the stem slightly oversize and plane down for a tight working fit. Plug the mortises with an offcut of stem when drilling the hole for the locking pin. The stem is cut away at the centre, firstly to take the cutter and secondly to take the brass plate. This latter is cut to shape, drilled for the screws, then rounded on its lower edge. Cutters for this tool should be of a standard width and thickness. Ideal material is a heavy gauge cabinet scraper from which pieces can be sawn and filed.

For work on a corner or well in from an edge, one of the two fences is simply removed.

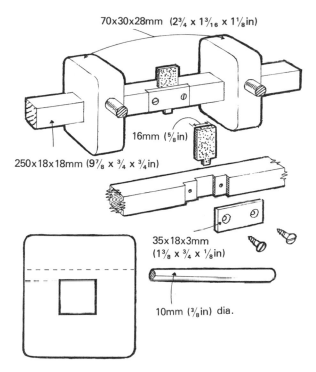

70x30x28mm (2¾ x 1³⁄₁₆ x 1⅛in)

16mm (⅝in)

250x18x18mm (9⅞ x ¾ x ¾in)

35x18x3mm (1⅜ x ¾ x ⅛in)

10mm (⅜in) dia.

Fig. 84

85 Cutter for inlay grooves

Straight line grooves for inlay stringings are normally fairly near an edge, so can be cut with a two-knife cutting gauge. The stock for this, preferably with a step, can be of any of the forms previously described. That shown at **A** is slotted and tightened with a thumb screw. It may be mortised or built up.

The stem **B** must be a good working fit in the stock. The end is mortised to take the two cutting gauge knives and their spacing material. An ideal material is printers' type spacing, easily obtained from a jobbing printer since so little is required. Convenient sizes are 12 point or 14 point. 12 point is ($\frac{1}{6}$in) wide, 14 point is a little larger. A piece of square section type, **C**, is called an em space. Half of that is called an en and below this are thick, middle and thin spaces. Combining a small handful of these will give micrometer adjustment between the knives. If type is not obtainable wooden blocks and veneer shims will have to do.

86 Circular groove cutter

For cutting circular grooves a similar stem or arm is required as shown, the mortised end is thickened up before the mortise is

Fig. 85

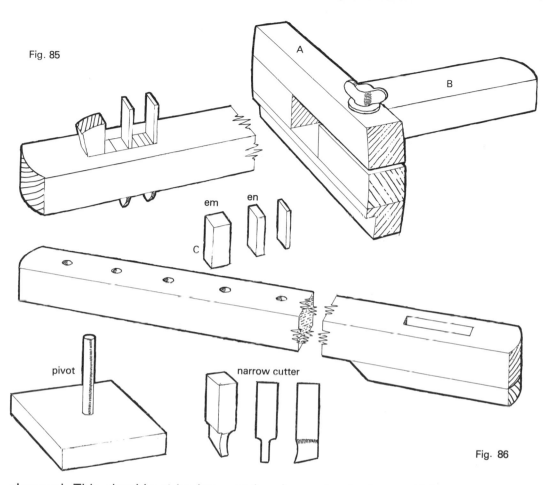

em   en

C

pivot   narrow cutter

Fig. 86

chopped. This should not be too short. From behind the mortise a series of holes is drilled. The spacing must be less than the mortise length. The pivot is made by fitting a long brass woodscrew to a wood block. Size is not vital – 8 or 10 gauge would suit but the holes in the stem must suit the screw. The head is countersunk on the underside and the thread sawn off leaving only the plain shank. The pivot block is glued to the work, with a paper joint

and using animal glues which can be soaked off later. The addition to the stem at the mortise must match the thickness of the pivot block. A small chisel will pick out the waste.

When a very narrow groove is required a cutter must be ground or filed to size from a small piece of square tool sheet.

Fig. 87

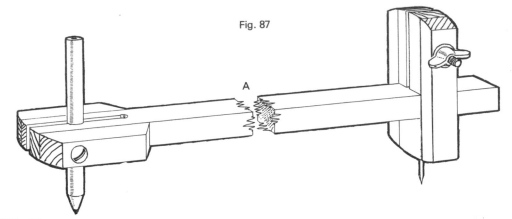

A

87 Beam compasses

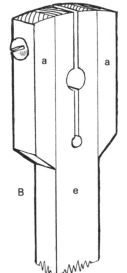

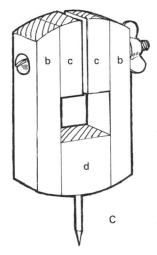

This excellent and virtually cost free tool is shown assembled at **A**. In constructing it, first machine an overlength piece of square section material, say 16 × 16mm ($\frac{5}{8}$ × $\frac{5}{8}$in). Cut off five pieces to make respectively, pieces **a**, **b**, **c** and **d**. All except **d** are sawn in half. The inside ends of **c** and **d** are finished quite square and all other ends are angled. Glue the **a**'s to the stem and **c** and **d** between the **b**'s using a short waxed block cut from the stem as a spacer. Hold the pieces together flat on a piece of polythene sheet. Drill a hole in end **B** for a pencil or ballpoint, the latter is often better, and then drill a small terminal hole of 3mm ($\frac{1}{8}$in) and a lateral hole for the clamp screw. A brass roundhead 12ga × 1$\frac{1}{2}$in is well suited for the cramp up. Drill half way at 6mm ($\frac{1}{4}$in). Drill the remaining distance at 3mm ($\frac{1}{8}$in). Saw the slot with a saw having a wide kerf. Screw up and test with the chosen pen or pencil. File off any protruding screw point.

A similar routine is adopted for the sliding point unit. There are several possibilities for the clamping screw. Either put a clear hole halfway through and tap the other half $\frac{1}{4}$in BSW, or metric equivalent. Use a thumbscrew to tighten or make one by soldering a wing nut on to a piece of screwed rod. Or solder a wing nut to a brass woodscrew. Screw in a normal woodscrew first then replace with the one made up. Or drill clear holes right through and use a 5mm ($\frac{1}{4}$in) coach bolt with wing nut.

The point can be made by grinding up a piece of silver steel of about 3mm ($\frac{3}{32}$in) diameter. Clean up the whole job, lightly sand and finish either with a poly-urethane varnish or teak oil.

87 Beam compasses, alternative design

The beam is a length of dowel 8mm ($\frac{5}{16}$in) or, better, 10mm ($\frac{3}{8}$in) diameter. A block is made for the point, a piece of 3mm ($\frac{1}{8}$in) silver steel, and similar blocks are made for pencil and ballpoint work. Woodscrews are quite adequate to clamp the blocks on to dowel and ballpoint. In cases where the centre point mark must not show, glue on a small wood block with a paper joint and pivot from this.

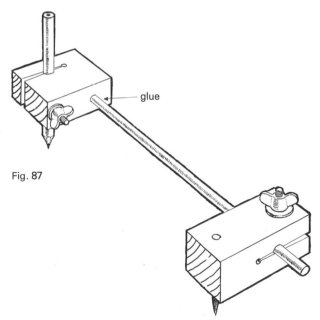

glue

Fig. 87

88 Spokeshave blade holders

The modern spokeshave blade is much too small to be hand held for sharpening. Two convenient holders are illustrated. The first, **A**, is of wood and is quite straightforward. The wing nut tightener is an improvement over a mere slot. Two small pins prevent the blade going in too far. They are in loose holes in the top and are driven into the bottom.

The second, **B**, is of metal, 3mm ($\frac{1}{8}$in) is quite suitable. A cutter is cramped in place and two holes are drilled through for the pegs which locate the cutter, which is held in place by the normal spokeshave wedge. A shaped handle can be filed up or a tang made to be driven into a turned wooden handle.

When a horizontal electric grinder is available the parallel sided version, **C**, is preferable as this will clamp into the tool holder in the manner of a plane iron.

89 Laminated mallets

The standard, commercially-produced mallet is not to every workers liking. Often the material is soft, wearing hollow very quickly. Handles may be unduly long and sometimes taper away from the head — the opposite to what is required. Cutting the deep tapered mortise, however,

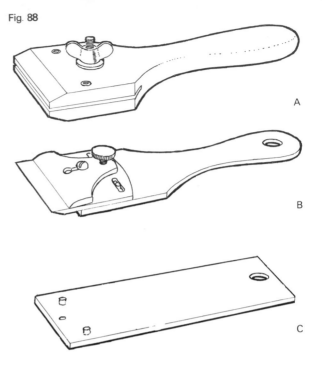

Fig. 88

A

B

C

66

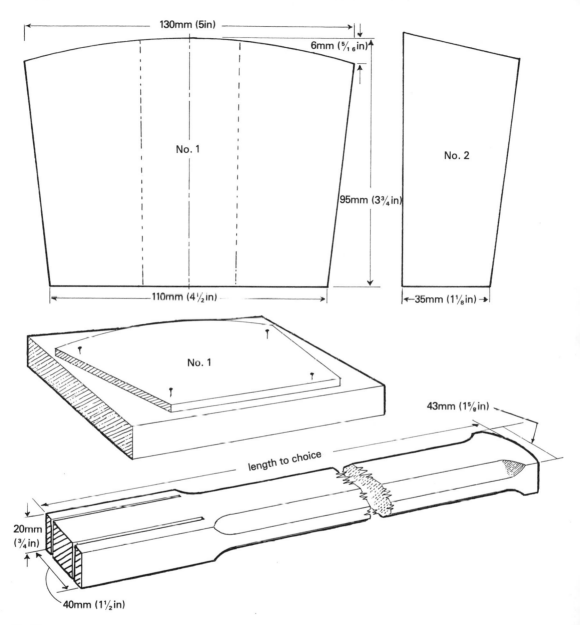

130mm (5in)

6mm (⁵⁄₁₆in)

No. 1

No. 2

95mm (3¾in)

110mm (4½in)

35mm (1⅛in)

No. 1

43mm (1⅝in)

length to choice

20mm (¾in)

40mm (1½in)

Fig. 89

is what prevents most workers from making their own.

Good mallets for normal or special use can easily be made by lamination, the pieces being cut to size on the sawbench, using the template fence described on page 146. Two templates are required made from 6mm ($\frac{1}{4}$in) ply. The faces of the template should be marked clearly **A** and **B** respectively. Any good hardwood is suitable; beech, oak, elm, ash, walnut and tropical timbers have been used. This should be machined to one third of the desired mallet thickness, 20 to 25mm ($\frac{3}{4}$ to 1in) or even greater. The components are cut roughly over-size. Template 1 is pinned on, face **A** up and cut out using the template fence Fig. 160C. To cover slight inaccuracies this is repeated with face **B** up. Two pieces of template 2 are cut and one handle. Ideally this should be ash or hickory. The mallet head pieces are now glued together with a good, freshly-mixed, synthetic resin glue. They may be pinned with brass or bronze nails or dowelled to resist the tendency to slide, but this is not essential, being largely a matter of taste. While in the cramps wash out any glue from the mortise.

If the gluing up has been carefully done the minimum of cleaning up of the head will be necessary. One or two shavings from the handle will permit it to enter

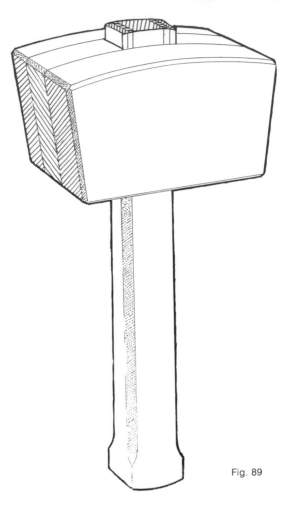

Fig. 89

the mortise but it will be found very loose at the top. Two carefully marked out and sawn wedges will secure the head. Slight shaping of the handle for comfort is best done after wedging with the head held in the vice and a slight chamfer round the head prevents the corners from splitting. A coat or two of varnish keeps the tool looking smart.

Out of a large number of such mallets no examples have been found with glue failure. Very careful planing, preferably by machine, and good fresh glue have ensured this.

90 Carvers' mallets

These mallets are not always easy to buy but are quite easy to make. Some workers prefer them for normal bench work as well as for carving. Many hardwoods are suitable for the heads, especially fruit woods. Small trees felled in the garden are particularly useful. The handle is short so almost any wood will do for this. Two sizes are shown and two common shapes. There are, of course, any number of variations. Heads are normally turned between centres but it is common to find the grain at right-angles to the shaft. The shaft is wedged but not glued.

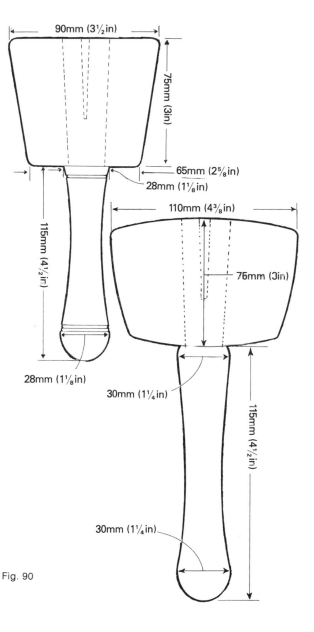

Fig. 90

91 Handscrews

These tools are much less common than they were a generation ago. Nevertheless they have several advantages over the much more numerous G cramps. They are lighter in weight, they do not damage the work and, of course, they can be made. Jaw length can vary between 300mm (12in) and 100mm (4in). They are usually square in section and are made from any close-grained hardwood. The screws can be of wood, if a wood screw box is available, or can be of bought metal screwed rod. The latter would be of a smaller size. The metal screws can be screwed, glued and even pinned into chisel-type handles and wood screws can be similarly fitted if it is required to cut out some of the woodturning.

It must be stressed that both the threaded holes are in the same jaw, in this illustration the lower one. In use the through screw makes the preliminary grip then the second shorter one screws into a cavity in the upper jaw thus increasing the pressure.

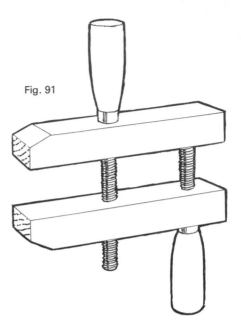

Fig. 91

92 Light cramps of fixed sizes

Fig. 92

The weight of ironmongery on relatively small jobs is sometimes out of all proportion and in a home workshop or in a busy communal workshop a large number of cramps may not be available. These cramps are an attempt to solve the problem of weight and number. A typical use, glueing in drawer slip, is illustrated.

Any hardwood will do — beech being preferred. Small offcuts can be used up and a small saw-bench and drilling machine make light of the task. Coachbolts and wing nuts are preferred but is may be difficult to find coachbolts in imperial sizes or wing nuts in metric. In the absence of wing nuts, hexagon nuts and a tubular box spanner will have to be used. Beware of adding too much weight, 8 mm ($\frac{5}{16}$in) or even 6mm ($\frac{1}{4}$in) coachbolts are quite strong enough.

Face the jaws with adhesive plastic tape. This stops glue from sticking and is easily removed. The hole in the moving jaw should be slightly elongated with a rasp to give easy operation. The square hole for the coachbolt head will have to be cut. In a hardwood it will not be possible to force in the bolt.

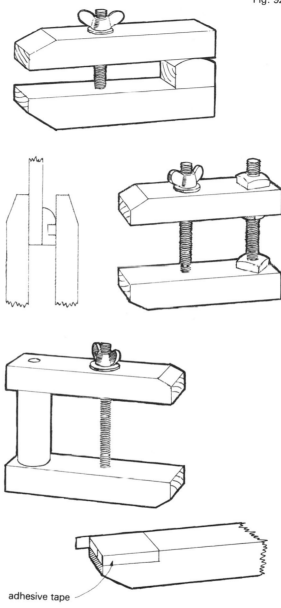

adhesive tape

93 Light board cramps

The tool detailed at **A** was originally designed to be used as in **B** to prevent spelching or splitting off when planing end grain. It is lighter, more convenient and more effective than struggling with a heavy sash cramp. The hardwood block is slightly drilled to take the screw end. A refinement is to put a steel disc at the bottom of the hole to prevent wear.

Light boards can also be glued up, using two or three of these cramps as in **C**. Lengths can vary to suit the work most commonly done and 20mm (¾in) and 25mm (1in) capacities are probably the most convenient. Suggested dimensions are shown at **D**.

94 Bench holdfast

This is built on the same principle as the handscrew, Fig. 91, and operates through a very small hole in the benchtop, compared with the iron holdfasts, so can be arranged to work almost anywhere. A long coachbolt, or a fabricated equivalent, runs through the hardwood jaw, through the bench top to the adjusting wing nut with large washer. The clamping screw, secured in its wooden chisel-type handle, passes through a square nut let into the underside of the jaw. The bench top is protected, either by a hardwood or metal

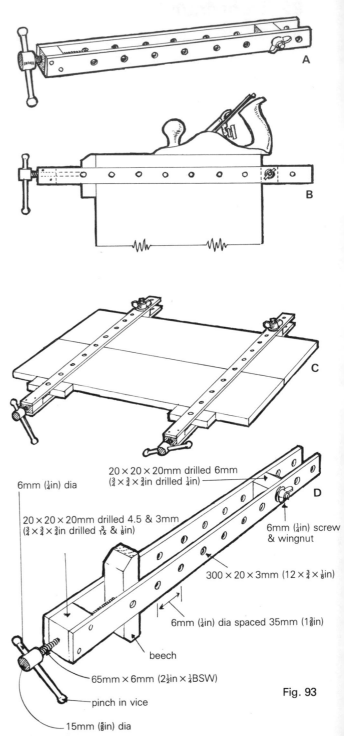

A

B

C

6mm (¼in) dia

20 × 20 × 20mm drilled 6mm
(¾ × ¾ × ¾in drilled ¼in)

D

20 × 20 × 20mm drilled 4.5 & 3mm
(¾ × ¾ × ¾in drilled ³⁄₁₆ & ⅛in)

6mm (¼in) screw
& wingnut

300 × 20 × 3mm (12 × ¾ × ⅛in)

6mm (¼in) dia spaced 35mm (1⅜in)

beech

65mm × 6mm (2½in × ¼BSW)

pinch in vice

15mm (⅝in) dia

Fig. 93

pad or else by a properly-made cramp foot. Average sizes would be 10 or 12mm ($\frac{3}{8}$ or $\frac{1}{2}$in) diameter screws and 35 × 28mm (1$\frac{3}{8}$ × 1$\frac{1}{8}$in) for the jaw.

One or two of these holdfasts are particularly suitable for holding low relief carvings or work for power routing.

95 Small job vice

This is the woodworker's version of the engineer's hand vice. It is held in the bench vice and proves useful for all sorts of small jobs, particularly wooden jewelry, pattern and model making, as in **A**.

Sizes will depend on the type of work contemplated and it should be constructed from any dense hardwood such as beech. Two pivot plates of plywood or, better still, of brass are screwed at one end to the moving jaw and to keep the jaws roughly parallel a number of pivot holes are provided. The pivot can be a length of 3mm ($\frac{1}{8}$in) diameter rod, cranked at one end for a grip. The jaws can be lined with wood, leather or rubber. The fixed jaw is longer and is gripped in the bench vice. A coachbolt is fitted into this, and is provided with a wing nut and large washer. To cope with the jaw movement, the hole in the moving jaw is extended by slight filing into a slot. The vice can be made in varying widths, the narrower one being useful for fine work.

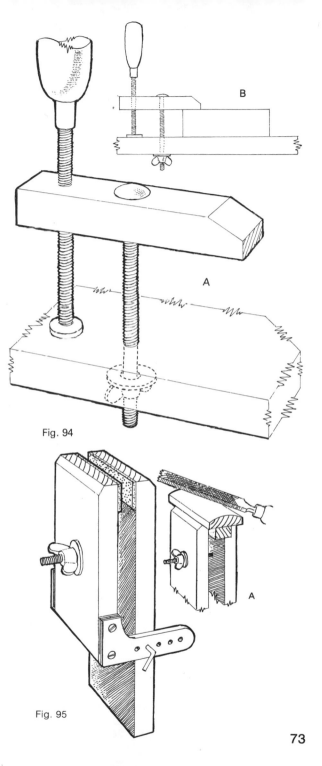

Fig. 94

Fig. 95

96 Vice clamps for sculpture

Fig. 96

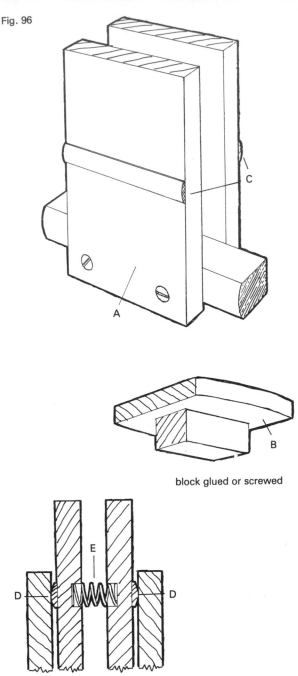

The holding of sculpture, particularly of the flat back variety such as plaques, presents a holding problem if a sculptor's vice is not available. The device drawn at **A** is made of two jaws or chops of beech or similar hardwood of about 20mm (¾in). A block of about 25mm (1in) square and long enough to fit over the slide bars of the bench vice is glued to the bottom of one jaw. Its opposite face is slightly rounded and the other jaw is secured to it with well countersunk screws in overlarge holes, permitting some movement.

The block screwed or glued to the underside of the sculpture, **B**, should also be about 25mm (1in) thick and if glued, use a scotch (animal) glue with a paper joint. Rest the jaws of the clamp on the bench vice bars and in this position glue on two small strips to take the pressure near the top of the vice jaws, **C** and **D**. A spring, **E**, located in two shallow holes permits the jaws to open when the vice opens.

F shows a removable high vice jaw which fits over the normal vice jaw. 20mm (¾in) beech screwed to 13mm (½in) plywood is suitable for this pair of jaws. Assuming that 25mm (1in) blocks are fixed to the work, then a spacing block of slightly less

block glued or screwed

bench vice jaws

74

than this is needed at the bottom of these jaws. It should be long enough to rest on the vice bars without falling off or through.

97 Plough plane – fence modification

Beginners find the plough plane far from easy to use and the further the groove is cut from the edge the more the difficulty increases. This arises from the blade not being kept upright. It is common practice to fit a wooden fence to the metal fence whose increased depth helps. Further help towards an almost foolproof method can be gained by arranging a rebate in the wooden fence which naturally prevents the tool from tipping. The rebated fence will have to be repositioned for each change in depth of cut. If much ploughing is anticipated an adjustable stop can be fitted to the wooden fence.

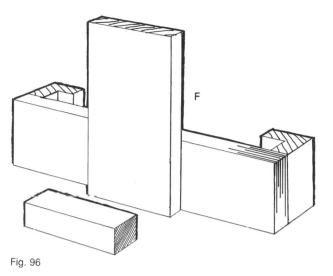

Fig. 96

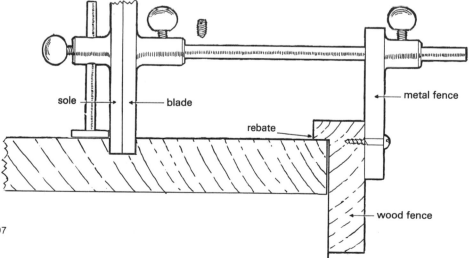

sole — blade — rebate — metal fence — wood fence

Fig. 97

98 Improved shooting board

The conventional shooting board suffers from several disadvantages which have been eliminated in this model. Beginners and the inexperienced are liable to cant over the plane cutting away part of the board and the planing stop, thus causing inaccuracies. Working against the planing stop (a very old fashioned and inefficient device anyway) does not give the stability obtained here by gripping in the bench vice. Once the plane has cut the smallest of rebates in the guides no further damage to the board is possible.

A packing piece of appropriate thickness makes sure that the cut takes place approximately in the centre of the blade. This packing piece can be replaced by false tables arranging for angles other than 90° to be cut. This board can be planned as left or right-handed and to take either the 50mm or 60mm (2in or 2⅜in) jack plane.

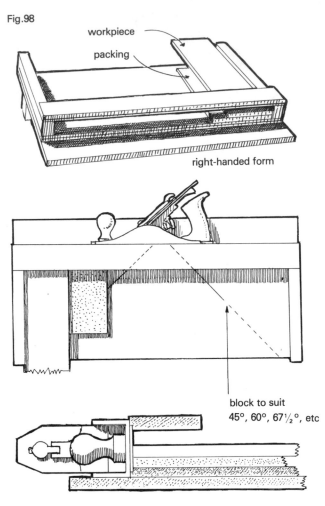

Fig.98

workpiece

packing

right-handed form

block to suit
45°, 60°, 67½°, etc

99 False tables for shooting board

The packing piece of the shooting board can be replaced by a false table, **A**, fitted with an angled block at 45° for mitring angles of rectangles, or 60° for the angles of hexagons, or $67\frac{1}{2}°$ for the angles of octagons. An alternative false table can be fitted with an adjustable stop, **B**. As the curved corner of this does not give efficient support to the workpiece it is necessary to interpose a parallel strip of hardwood.

100 False tables for angles other than 90°

Bevelling and edge jointing at angles other than 90°, as in coopered doors or stave-built turnery, can easily be achieved by false tables fitted to the basic shooting board. A number of these can be made from multi-ply or even chipboard supported by differing pairs of wedges. Drill a clearance hole through both the table and the wedge. Glue the wedge to the false table only and when tightening up the screw make sure that both the wedges and the table are pressed firmly against the sole of a plane, which is held in the working position.

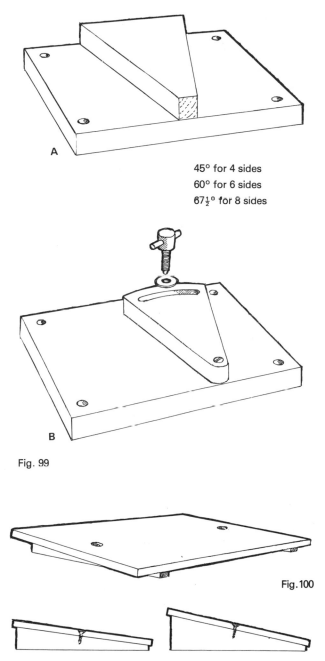

45° for 4 sides
60° for 6 sides
$67\frac{1}{2}°$ for 8 sides

Fig. 99

Fig. 100

101 Improved mitre shooting board

This tool is an alternative to the traditional 'donkey's ear' shooting board. It offers the same advantages as the previous board. It can be firmly held, it cannot be damaged and retains its accuracy. Additionally the workpiece can be more conveniently held horizontally than sloping upwards at 45° as with the traditional board. Left or right-handed models can be constructed to suit a 50mm or 60mm (2in or 2⅜in) plane.

102 Shaped plane handle

The use of the plane on the shooting board is made easier by the use of a side handle. The side of the plane, left or right, can be tapped to take either a turned knob or a carved handle.

103

Either of the two shooting boards described may be fitted with a slotted fence as illustrated. This is most convenient for planing thin material truly parallel either at 90° or 45°.

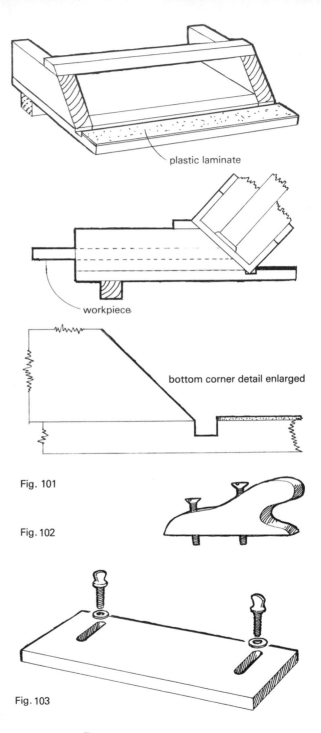

plastic laminate

workpiece

bottom corner detail enlarged

Fig. 101

Fig. 102

Fig. 103

104 Mitre blocks and boxes

Fig. 104

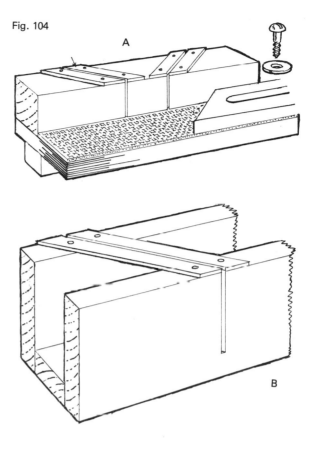

These do not as a rule have a very long life, particularly if used by many people and these improvements will prolong both life and accuracy. In **A** the block is made of a good hardwood glued to a multi-ply base. Beneath this is a stout strip with which to grip the block in the vice. The 45° markings are best made using either a combination square or a good set square. Saw carefully with a fine saw, then with the same saw in the kerf glue and pin on some ready-drilled strips of plastic laminate. In use, always feed in the saw from the front and never from the top. Protect the plywood base with a hardboard offcut. A slotted length stop cut to 45° at one end and 90° at the other is a useful accessory when a number of equal lengths are to be cut. For lengths greater than that of the mitre block cramp a stop to the workbench top.

Wider components can be better mitred in a box, **B**. Here again saw carefully, leave the saw in the slot and secure the plastic laminate guides which, as well as providing accuracy, also strengthen the mitre box. The hardboard strip and vice grip are equally needed.

105 A tool for mitred dovetails

The mitre saw cut on a dovetail needs to be cut very accurately the first time. Any paring back to fit will either leave a gap or necessitate cutting back the dovetails. This device is simply a hardwood block from 50 × 50mm (2 × 2in) material which has been grooved to take a piece of 13mm ($\frac{1}{2}$in) multi-ply which should project between 50–70mm (2–3in). Carefully saw the mitres with a thin saw and fit plastic laminate guides as described in Fig. 104.

To use, first cut the dovetails then position the joint carefully on the plywood beneath the saw slot. Small workpieces and the aid can both be gripped in the bench vice. Wider pieces, as in carcase work, should be held in the vice with the mitre aid cramped in place. Do not insert the saw from the top.

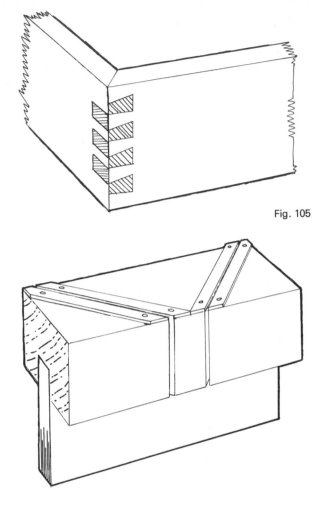

Fig. 105

106 Hand planing thin strips

Thin strips of identical thickness, such as may be required for laminating, can be accurately produced by hand planing by means of a simple jig. This consists of a base-block, **A**, and two rebated side members, **B**. The space between the two rebates must just allow free movement of the chosen jack plane. **A** projects below **B**, to be held in the vice.

80

The sides are glued and pinned in place using an assembly block with a true face in the plane position and a piece of ply, card or suitable spacing material of the required thickness. The illustration makes this clear. When complete, an end stop, **C**, is fitted.

Modifications For the making of stringings for inlaying or musical instrument making, grooves are ploughed or cut on the circular saw in the baseblock **A**. In this case there is no need for rebated sides. Very thin pieces will tend to buckle when planed against a stop. This is overcome by cutting away some of the baseblock and pinning on the workpiece below the level of the blade. In this case, of course, the components and the jig must be made extra long. An adjustable model can be made by slotting and screwing on the sides. The adjustment is made using the same method as when gluing on the sides to the simple model. Solid wood keys for reinforcing mitre joints, as described on page 24, can be produced in this manner.

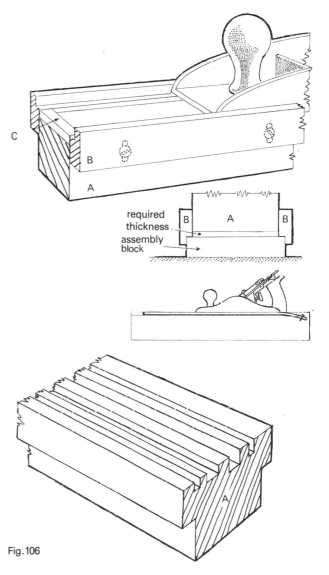

Fig. 106

107 Hand planing very small components

Very small components can best be planed by holding a plane upside down in the vice and pushing the workpiece over the blade. As this method gives every chance of shaving off the finger tips, a push stick is an advantage. Even better is this simple planing device. It consists of a hardwood base with a firmly secured handle. Guide pieces, thinner than the finished job, can be pinned or glued on so that they can be changed when the aid is used for another job.

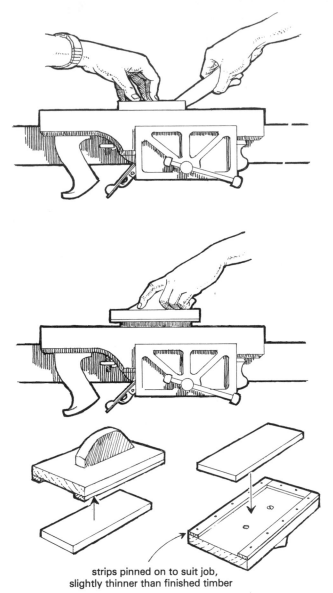

strips pinned on to suit job,
slightly thinner than finished timber

Fig. 107

108 Plane for plastic laminates

Although the normal bench plane can cope briefly with plastic laminates, a cutter ground to 25° and sharpened at 30° very quickly loses its edge. A more successful cutter can be made using a much less acute angle as at **A**. Grind the cutter in the normal way then turn it over and grind and hone the other side at 45°; this gives a section as in **B**. To prevent clogging, the cap iron should be sawn off at the end of the curve and filed smooth. The cap iron now serves only to move the blade. The other components remain unaltered. Assemble and use in the normal way. A little more pressure than usual will be necessary but not too heavy a cut should be used. This form of cutter will last much longer than the standard cutter, when used on laminate edges.

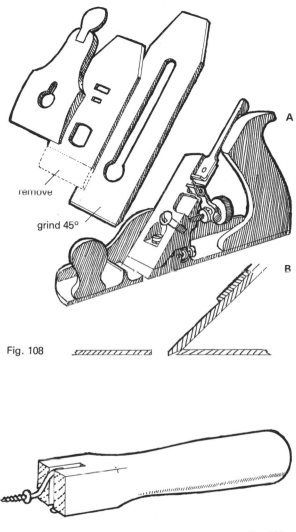

remove

grind 45°

A

B

Fig. 108

109 The hookdriver

Screwing a large number of hooks into correctly drilled pilot holes is certainly hard on the fingers. The use of pliers roughens the hook or, in the case of brassed or chromed hooks, damages the plating. The hookdriver made from any dense hardwood such as beech enables hooks to be put in in quantity and at speed with the minimum of effort and no discomfort.

Fig. 109

110 Wooden jack plane

Fig. 110

All the wood planes that follow are made by this built-up method, though they could be made in the solid if preferred. It is invariably advised that the end grain of the plane should look as shown at **A**, the medullary rays as nearly vertical as possible to prevent distortion. At the same time it has to be admitted that many old and perfectly satisfactory planes can be found in which this has been disregarded.

Wood Beech is the commonly accepted wood, preferably the red varieties. Hornbeam, maple, pear and cherry are common in Europe; boxwood and walnut are satisfactory though expensive. It would appear, then, that almost any good close-grained hardwood may be used.

Construction The centre block is first prepared accurately, generally 4mm ($\frac{1}{8}$in) thicker than the width of the intended blade. Follow the plans for the other dimensions. Length is kept a bit long. Make sure that the two sides are perfectly true, flat and parallel, and at right-angles to the sole. Prepare another piece the same length, of about 25mm ('1in) thickness and of the width of the centre block plus 4mm ($\frac{1}{8}$in) or so. Plane both faces with great care and mark both with a face mark. From each face

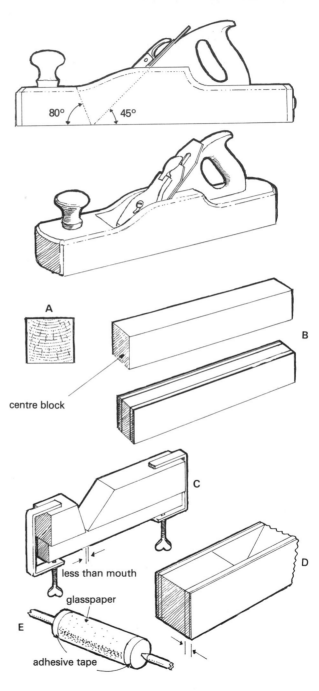

80°   45°

A

centre block

B

C

less than mouth

D

glasspaper

E

adhesive tape

84

gauge the thickness of the side pieces, **B**.

Next mark out the escapement and the mouth and cut the centre block into two pieces. This can be done by hand or on the circular saw using suitable angle blocks on the crosscut slide. A planer saw is recommended for a first-class finish. If it is intended to make the plane fully adjustable, a suitable recess for the mechanism should be cut at this stage. Not only is it more inconvenient later, but before the glue-up a router can clean up the recess nicely. Fig. 118 shows a variety of wedges and wedge fixing methods.

Assembly An assembling block made from any wood must be prepared. This should be as long as the plane and about 50mm (2in) wide and 13mm ($\frac{1}{2}$in) less than the thickness of the centre block. The assembling block is held in the vice and the two centre blocks are cramped to it, **C**. Space these to give a gap slightly less than the mouth required since the mouth will open when the sole is trued.

Saw off the two side pieces from **B** about 2mm ($\frac{1}{16}$in) on the waste side of the gauge lines. Do *not* plane to the line. This waste avoids the use of a lot of inconvenient cramping blocks. With the waste edge outwards glue up at once to the centre block before

any warping of the thin side pieces can take place, using a synthetic resin glue. A large number of G-cramps will be required to get a perfect joint. Nipping in the bench vice will certainly not do. Position the sides so that they project a little beyond the sole, their width allows for this. The arrangement is shown at **D** with cramps omitted for clarity. Make sure of a perfect contact between the pieces at all points. Once several cramps are on, the assembling block can be released and two further cramps added to the job. Leave for twenty-four hours, then remove the cramps.

Plane down the sides to their gauge lines and clean off the steps on the sole and the top. The sole is accurately trued later. Mark out and saw the side shapes, cleaning up with gouge, spokeshave and rasp. If a lathe is available, finish on a small drum sander, shown at **E**. The wooden drum is turned to about 50mm (2in) diameter. Glass-paper, preferably open coat, is cut to size and held on with adhesive tape. It is rather a bumpy ride but gives a first-class finish. Shape the ends and bevel the corners.

Handle and knob Now prepare the block for the handle. If the open type is required, it might be possible to incorporate a manufactured handle. This will be out of the question if the closed type as illustrated is chosen since the ramp on which the blade beds is continued up the handle for more support. Cut the mortise and fit the block to it. Shape the handle according to preference, having due regard to accommodating the mechanism if it is to be fitted. The closed handles are fitted with a slight projection which can be cleaned off with a small block plane so that the blade fits snugly.

The design for the front knob can be evolved to suit individual requirements, glue the knob in as the last phase of the whole job.

Cutter If the cutting units are standard components as used in Record and Stanley planes, the vital factor in fitting this unit is to position the lever cap screw exactly. To do this bind the cap iron and the cutter together in their working position with adhesive tape and remove the cap iron screw. Place the lever cap in its working position and bind this also. Stand the plane on a flat surface and insert the blade. Through the hole in the lever cap scribe the centre for the lever cap screw. If there is to be no adjustment mechanism remove the lever cap only, replace the cap iron and cutter and, through the hole for the cap iron screw, mark its centre. On this centre, with a forstner bit for preference, bore a hole to accommodate the cap iron screw. The lever cap screw may be a No. 12 round-head steel wood screw. An improvement is to make a brass bush from a short length of $\frac{1}{2}$in Whitworth screwed rod. This can be bored centrally with a $\frac{3}{16}$in drill, then tapped to take a $\frac{1}{4}$in machine screw with round head. Equivalent metric threads are equally suitable.

Finishing The most satisfactory method is to clean up the plane and give a few coats of either clear cellulose or clear french polish thinned 50/50. Rub off the nibs with waxed steel wool. True up the sole with the cutting unit in place, cutter withdrawn. Ensure that the lever cap screw is tight enough. If there is no mechanism the screw must be tighter than is the case in an iron plane. Test with straight-edge and winding strips. Wipe the sole periodically with raw linseed oil.

Adjustment by hammer is apt to damage a nicely-built plane, so where possible it is advisable to fit a striking button. This is equally effective whether placed on top at the front, or on the end at the rear. Boxwood or rosewood are suitable woods.

Test the completed plane on mild timber first, making sure to try it on a wide face as well as on a narrow edge. Keep the mouth as fine as possible but remedy clogging by gently paring or filing the mouth wider. An adjustable mouth, which can be fitted to the existing plane or incorporated into a future one, is described in Fig. 111; also several varieties of mechanism, all of which can be made with basic hand tools.

111 Jack plane with simple mechanism

This plane is built up in the same way as the non-adjustable version. The two centre blocks **A** and **B** are glued between side blocks. Before gluing, the front block, **B**, has a recess cut out to take the sliding mouth, **C**. The main block is passed over the circular saw to give a groove to take the adjusting arm, **H**. This is joined again on top by a capping piece, **F**. This piece is set back slightly from the bed to make room for the lateral lever **G**. This is identical to the lateral levers on the metal planes. The greater length of this plane makes it possible to fit the simpler mechanism. The adjustment screw, **J**, and wheel, **K**, are placed as far forward as possible, if necessary by cutting a slight recess in the block. M6 or ¼in BSF are suitable threads. If left-hand threads can be cut this will preserve the convention of

Fig. 111

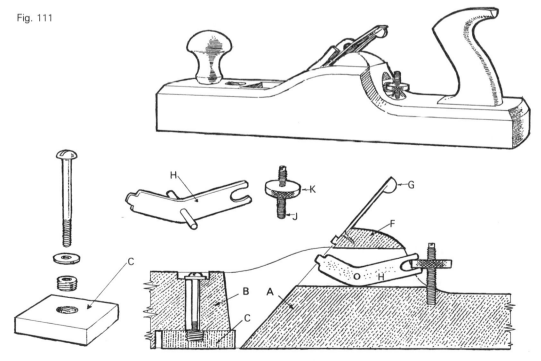

clockwise rotation increasing the cut. A cardboard template of the adjusting arm **H** should be carefully drawn and cut out. Drill the pivot hole 3mm ($\frac{1}{8}$in) and try out the pattern, correcting where necessary. Using the template make the arm itself using 4 or 5mm ($\frac{3}{16}$in) mild steel. Fit the arm, put in the lever cap, screw, assemble all the parts and check that the mechanism works properly.

The sliding mouth, **C**, has a metal insert or nut fitted to it. This may be screwed in or glued with epoxy resin. The locking screw operates through a slotted hole and holds the mouth firmly in place. When this operates successfully, screw it firmly in place and plane the sole true. Lastly fit the handles.

112 Adjustable wood smoothing plane

This is a delightful plane to use. Many readers will find the rear handle is well worth the work it involves. The 50mm (2in) size seems the most satisfactory, though a 45mm (1$\frac{3}{4}$in) could well be made to satisfy some special need. The use of a 60mm (2$\frac{3}{8}$in) cutter would produce a massive plane of doubtful advantage. The plane is fitted with an effective adjustment mechanism which can easily be made at home.

A closed handle is fitted and, as no satisfactory handle is manufactured, one will have to be made. Try to finish it at 25mm (1in). A thin handle, out of stuff 25mm (1in) sawn, is unpleasant to hold. Make a full-size drawing of the handle on cardboard which can then be cut out and used as a template. A similar template may be found useful for the body. The home-made sanding drum described on page 84 is very helpful used on the body but there is no short cut to making the handle. Any means are justifiable and they may include gouges, rasps, files and narrow scrapers made from hacksaw blades. These will eventually produce a shape that suits. Before fixing it, do not forget to cut back the handle to accommodate the lateral lever and groove it to take the spindle.

The body is built up with two sides glued to the two centre blocks in much the same way that the jack plane described on page 84 and, in fact, a similar assembly jig can be rigged up. Resin glue should be used. Whatever finish has been chosen, give several coats before gluing in the handle. This prevents the glue, which oozes from the handle joint, from sticking to the surface.

The adjustment mechanism follows. In making it ensure that the length of the spindle and the lateral lever is such that the plane will take a new full-length cutter. According to its size and style, the top of the handle may have to be modified to accommodate the brass knob.

Mechanism This can be made with hand tools by any woodworker of average skill. It can be fitted to this plane, or it can be added to an existing plane. The whole device is made from bright mild steel, except for its brass knob. The assembled parts seen from the side and above are shown at **A**. The individual pieces at **B**. The base plate consists of a strip 20mm ($\frac{3}{4}$in) by 4mm ($\frac{1}{8}$in) approximately 75mm (3in) long. From a length of 15 × 15mm ($\frac{1}{2}$ × $\frac{1}{2}$in) steel cut three blocks 20mm ($\frac{3}{4}$in) long. File one down to a thickness of just under 13mm ($\frac{7}{16}$in). This block and one other are fixed to the base strip 45mm (1$\frac{3}{4}$in) apart. They may be brazed, riveted with 4mm ($\frac{1}{8}$in) diameter rod or screwed from beneath. The plate is next bored and countersunk to take two No. 6 wood screws. Having marked out the centres, screw this assembly to a right-angled block, **C**, making sure that it is vertical. Cramp the centre moving block exactly beneath the top fixed block and, using a drill press or mechanically-held drill and 5mm ($\frac{3}{16}$in) bit, drill through them all. Remove the moving block and put a 6mm ($\frac{1}{4}$in) drill through the top block. Now reverse the whole and on

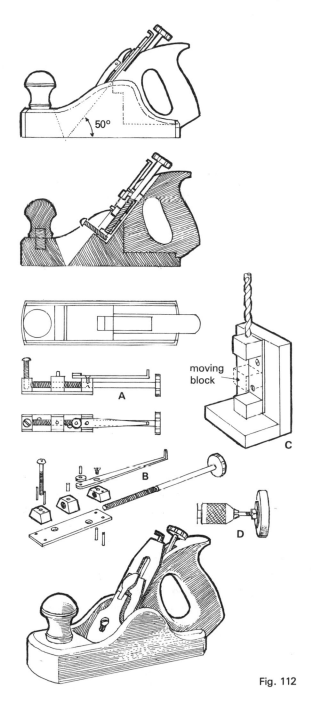

Fig. 112

the same axis drill the bottom block with any drill between 1·5mm ($\frac{1}{16}$in) and 3mm ($\frac{1}{8}$in). Reverse again and bore into the inner side of the lower block, on this pilot hole, 6mm ($\frac{1}{4}$in) deep using a 6mm ($\frac{1}{4}$in) drill.

The centre sliding block is drilled on top and tapped to take a 5mm ($\frac{3}{16}$in) diameter peg. With this firmly in position again clear the original 5mm ($\frac{3}{16}$in) hole and tap with $\frac{1}{4}$in BSF. Whitworth will do at a pinch but will not give so fine an adjustment. The normal thread will require an anti-clockwise turn to increase the cut. To obtain the conventional clockwise rotation a left-handed thread must be used. Left-handed taps and dies can be readily obtained from any reputable tool dealer.

To make the spindle, thread a 135mm (5in) length of $\frac{1}{4}$in diameter for 55mm (2$\frac{1}{8}$in). Take care if unpractised to start the thread squarely. Return to the body and on top of the upper block, drill and tap at $\frac{1}{8}$in to meet the $\frac{1}{4}$in through hole. Do not put through the plug tap but leave the thread tight. File a $\frac{1}{8}$in keyway round the spindle at the appropriate place. This can be done in the chuck of a woodturning lathe. Assemble and test for smooth working with a $\frac{1}{8}$in screw in place.

Make the lateral lever from a strip of 10mm × 1·5mm ($\frac{3}{8}$in × $\frac{1}{16}$in) stock. Drill a 3mm ($\frac{1}{8}$in)

diameter hole near one end for the rivet and another 13mm ($\frac{1}{2}$in) from it for the pivot screw. Countersink each appropriately. From a 12mm ($\frac{7}{16}$in) diameter rod, previously drilled at $\frac{1}{8}$in cut off a 1·5mm ($\frac{1}{16}$in) slice. Rivet on this stud, then assemble. File the end of the pivot screw until it holds the lever firm yet does not foul in the keyway. The lever cap securing screw hole is now drilled in the lower block. A $\frac{1}{4}$in BSF or Whitworth round-head screw will do the job. Now assemble in the plane with a full-length cutter to see where to cut off the spindle and crank the lateral lever. The position in the plane of the lever cap screw is of importance as this controls the position of the lever cap. If this comes too low the mouth will clog.

Make the knob from a brass disc 25 × 8mm (1 × $\frac{5}{16}$in) thick. Drill a 5mm ($\frac{3}{16}$in) hole in the centre and tap $\frac{1}{4}$in. Thread some $\frac{1}{4}$in diameter rod on a lock nut, and screw on the disc as at **D**. Turn up in the chuck of a woodturning lathe or electric drill. A hand turning tool can be ground from an old file; the cutting angle for brass is 90°. Twelve notches filed round the edge give an adequate grip. Thread the spindle end and screw on the knob and then wedge or pin it securely. Cut off the lateral lever 10mm ($\frac{3}{8}$in) past the bend and round its tip.

Cut off the peg at a suitable

length in the cap iron slot and ensure that the housing in the plane body is deep enough, and that the lateral lever does not foul on the wood anywhere. Adjust the lever cap screw to a convenient degree of tightness. Sharpen and test, starting on some mild softwood.

The adjustment mechanism could also be fitted to the jack plane.

113 Adjustable scraper plane

This type of plane has the advantage over the spokeshave type in that having a longer sole, it is not so liable to plane hollow. Also, being wooden, it slides more sweetly. Similar planes, called veneer planes and made of iron, were available in at least two models many years ago. The present plane, though fulfilling a similar purpose, is not a mere reproduction of them. The need for the wooden sole must have been felt before, since at least one model of scraper plane had provision for the attachment of a wooden sole. This plane can be made with only basic hand tools though it is a great help if the escapement is cut with a fine circular saw.

Centre stock Prepare as described on page 84 with a width of 63mm (2½in). Mark out and cut

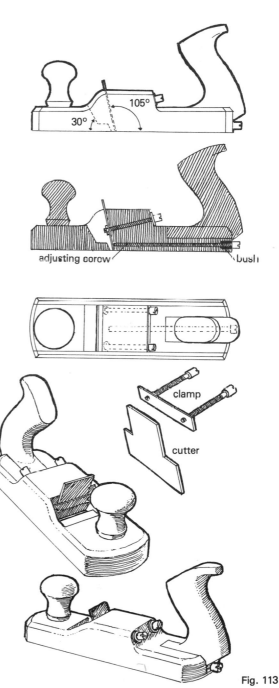

Fig. 113

the escapement. The angle on the rear section is a forward leaning angle of 105°; the front angle is 30°. A small step cut on the front block helps to keep the mouth small and at the same time helps to reduce the chance of shavings piling up behind the clamp bar and causing clogging.

Next bore the 6 mm ($\frac{1}{4}$in) hole for the long adjusting screw. Locate the centres and bore from each end. A lathe is a great help towards accuracy here. This screw can run through a brass bush as the drawing shows. It is made from a piece of $\frac{1}{2}$in Whitworth screwed brass rod. This is drilled centrally and tapped at $\frac{1}{4}$in. Alternatively a brass $\frac{1}{4}$in nut can be sunk into the bed behind the blade. The clamp screws can run in grooves cut in the centre block before the glue-up. Otherwise, they too may be bored if accurately marked at each end.

The sides are prepared in the usual manner and glued on as soon as possible after they have been sawn out to minimise the chance of their warping. Clean up and fit a manufactured handle. Glue it in place. Turn the knob according to individual taste and fix with glue. Make sure that it comfortably clears the blade.

Metal work The clamp bar is a piece of 15 × 3mm ($\frac{1}{2}$ × $\frac{1}{8}$in) mild steel, drilled and tapped $\frac{1}{4}$in Whitworth. All three screws are easily

made from brass by soldering $\frac{1}{4}$in screwed rod into a tube of an inside diameter of $\frac{1}{4}$in, all three need a sawcut for the screwdriver and the two clamp screws require a brass washer each.

The blade itself can either be a manufactured scraper blade, or made from a really heavy gauge cabinet scraper. The latter are particularly hard, but they can be filed and sawn; they are very hard on hacksaw blades. In some cases it may be worth softening the blade and subsequently re-hardening. Sharpen the blade with a file, and then oilstone it at 45° as described on page 59. Burnish over the edge and then stand the plane on a flat surface, drop in the blade and tighten up the clamps.

A fine cut will probably be obtained without the use of the adjusting screw and this is the method which should be aimed at, keeping the adjusting screw in reserve to increase the cut. The more the cutter is bent by the adjusting screw, the narrower will be the shavings. In operation, try to produce several fine shavings rather than a few thick ones. Cut parallel to the grain but hold at an angle to get a skew cut. Occasionally wipe the sole with a drop of raw linseed oil, otherwise maintenance is unnecessary.

114 Compass plane

Fig. 114

The compass plane, for working shallow curves, is built up in the same way as the other laminated planes. The model with a sole curved in length only is quite straightforward. A plane curved in width also, presents a slight problem. Such a plane might be used, for example, to clean up a dished chair seat. The curved end to the blade necessitates a curved mouth slot. This is best achieved after glue-up and before shaping by chopping out a recess in front of the blade and fitting in a block as for remouthing a plane. Keep the mouth extra fine and, after shaping the sole, carefully pare or file away the block to give a fine mouth matching the blade shape.

115 Toothing plane

This tool, once bordering on extinction is now increasingly used to roughen the ground not only in preparation for veneering but also for plastic laminates. This handled version is much more convenient to use than the coffin-shaped model. The laminated construction is the same as that of the planes described earlier. Any of the blade securing methods is suitable and a good tool dealer can still supply toothing blades. The pitch of the blade should be 85°.

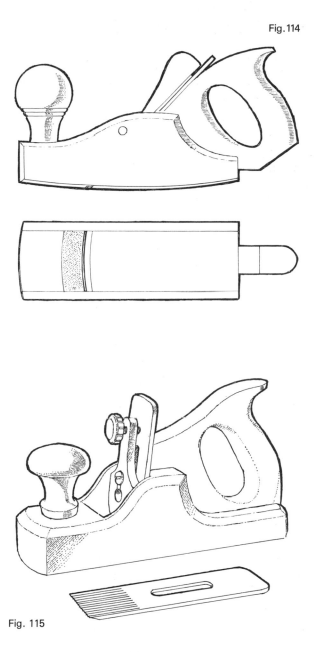

Fig. 115

116 Improved router

The addition of the levelling foot greatly increases the scope of this router by allowing it to work on such features as oversize tenons and rebates as shown in **A**. The basic construction from **B** is straightforward, it is made from 25mm (1in) beech or other dense hardwood. Cut out and bore the base then shape the block and glue it on. Drill the block for the clamp bolt 10mm (⅜in) then chop a shallow square recess of 13mm (½in) for the head. Cut a shallow housing of 6mm (¼in) in the base for the cutter. Drill the ends and glue in four nylon knock-down fittings. Drill holes of 16mm (⅝in) then turn the handles to suit. Varnish the body and handles with three coats of polyurethene varnish and then glue in the handles. The levelling foot, **D**, is simply a slotted hardwood block secured with two round-head screws with big washers at either end as required.

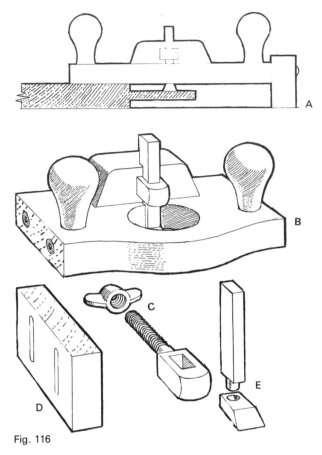

Fig. 116

The shank of the cutter, **E**, is made from 6mm (¼in) square mild steel with a round tenon filed on the end. The cutters themselves are filed from tool steel. Drill and countersink then rivet and braze together. Harden and then temper to light brown, see page 204.

The clamp bolt, **C**, is turned or filed from a piece of 13mm (½in) square mild steel and threaded 10mm (⅜in). Drill and file a hole for a loose fit on the cutter shank. A large wing nut and washer secure it through the block.

117 Shoulder plane

A satisfactory shoulder plane can be made with hand tools and a drill without great effort or skill. The two sides are roughly cut out from 3mm ($\frac{1}{8}$in) mild steel and two of the holes are drilled so that they can be bolted together and filed to shape. A sole, of 10 × 10mm ($\frac{3}{8}$ × $\frac{3}{8}$in) steel is held in position using the two bolts and a packing piece. The three components are then drilled, countersunk and riveted together. This is followed by the 6mm ($\frac{1}{4}$in) strip above the wedge. The wood filling, preferably of rosewood, is next inserted and riveted in place. The assembly is now filed up true, drawfiled and finished with emery. The blade (of 3mm ($\frac{1}{8}$in) or 2mm ($\frac{3}{32}$in) tool steel) is filed to shape and hardened and tempered. (See details on page 204.) The wedge completes the job. Give great care in filing up the mouth, which must stay as fine as possible. A suitable length is 125mm (5in).

Fig. 117

118 Alternative wedges for wooden planes

Patterns **A**, **B** and **C** can be made either from a dense hardwood into which has been let a brass nut, or by sawing out from a solid piece of metal (brass looks best), or from a simple aluminium casting which is well within the capacity of a school or college workshop. **A** pivots on a large round-head woodscrew as the cap screw, **B** notches on to a steel pin, permanently fixed into the plane sides and **C** is best as a casting. This pivots on a removable pin which passes through two brass pivots glued into the plane sides. One pivot is threaded, the other has a clear hole. **D** is a hammer-adjusted, wooden wedge operating against a pivoting wooden bar. To ensure a really precise fit of the wedge on the blade it is essential that the bar should not be glued but left to move freely.

119 Dowel cutters

Dowels are seldom used singly, one generally requires a number of identical sizes. Two devices are shown for hand sawing. The long holes should be a fraction over the nominal size of the dowel to give an easy fit.

Device **A** will cut four standard sizes. Grip it in the bench vice and feed in a length of dowel

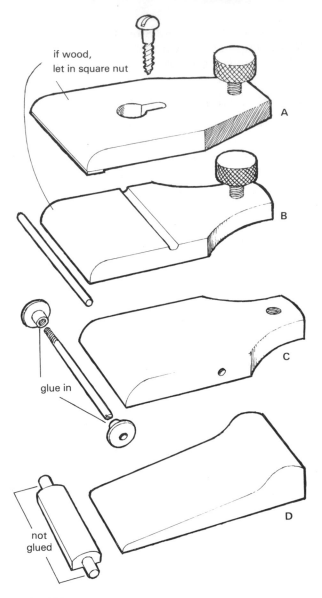

if wood,
let in square nut

A

B

glue in

C

not glued

D

Fig. 118

flush with one end and saw through at the appropriate sawcut. Device **B** is gripped by the rebates in the bench vice. A small pin of, say, 3mm ($\frac{1}{8}$in) diameter fits into holes at 10mm ($\frac{1}{2}$in) intervals and acts as a stop. After each cut the pin is removed and the cut dowel is pushed out by the length of dowelling. The pin is replaced and the next dowel cut.

The long holes, or small preparatory pilot holes, are best bored on the lathe or on the drilling machine using a spike on the table, see Fig. 156.

120 Dowel plate

Short lengths of dowel can easily be produced from the dowel plate. This not only obviates the purchasing of dowelling but enables dowels to be made of the same material as the rest of the job. The dowels made are quite satisfactory for joint work but long lengths to be used as rails and spindles should not be expected. The plate itself can be made from either mild steel or gauge plate tool steel of about 8mm ($\frac{5}{16}$in). The graded dowel holes are accurately drilled and, if possible, reamed to a slight taper and four screw holes drilled and countersunk. If made of tool steel the piece is hardened and tempered. If of mild steel it is case hardened. See details on page 204.

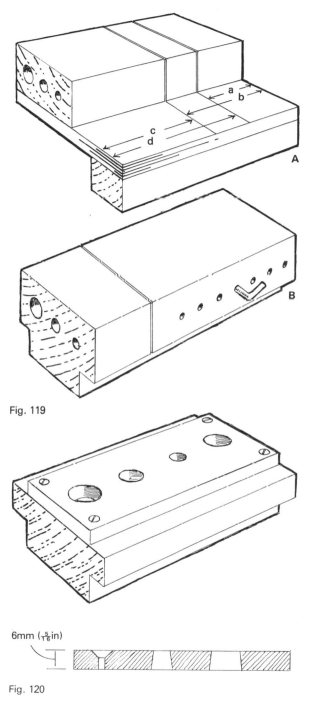

Fig. 119

6mm ($\frac{5}{16}$in)

Fig. 120

A hardwood baseblock is made with rebates which fit over the vice jaws. Holes are drilled to correspond with those in the plate but these are made larger to give clearance to the dowels. The plate is screwed to the base.

Fig. 121

To make a dowel, roughly plane the wood to an octagonal section before driving it through a hole of larger size than is required. Then use successively smaller holes until the finished size is obtained. Drive only with a mallet to avoid damage to the cutting corner and finish the drive if necessary with a wooden punch.

121 Dowel groover

Dowels require a groove to enable excess glue to escape. A hardwood block of about 15mm ($\frac{5}{8}$in) thickness is drilled to take the common sizes of dowels used. A gauge 8 steel woodscrew is arranged as shown for each hole and regulated to give the required size of groove. The block is held on the open vice and the dowels are knocked through, either with a mallet or with a hammer and a wooden punch.

122 Dowelling aids

A shows a tool intended for the dowelling of narrow components, for example legs and rails or narrow flat frames. For a few jobs, a hardwood block and a

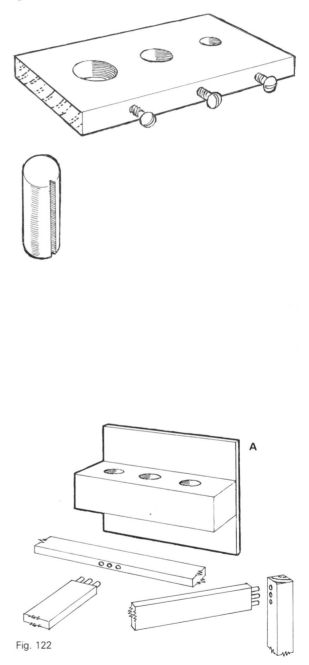

A

Fig. 122

98

plastic laminate or plywood fence is adequate. For a lot of use mild steel can be used. The bores can be case hardened.

The model at **B** is generally made up for a particular job. It is intended for jointing the wider members of carcase constructions either in wood or chipboard. Both must be firmly cramped to the work and the larger checked with a try square.

123 Saw filing aid

This simple aid helps to maintain a fixed angle when sharpening handsaws and also makes it easier to maintain an angle after a break in filing. A good handle is required preferably with a brass ferrule. A second ferrule fits snugly on this, but must turn freely. Into this is screwed a metal rod of about 75mm (3in). If the ferrule is thin it is worth silver soldering on a brass nut to increase the thread.

To use, set up the saw in a saw vice and press down the file firmly into a good gullet, usually near the handle. Now adjust the rod, to preference, either vertical or horizontal, and this position can be fairly accurately maintained.

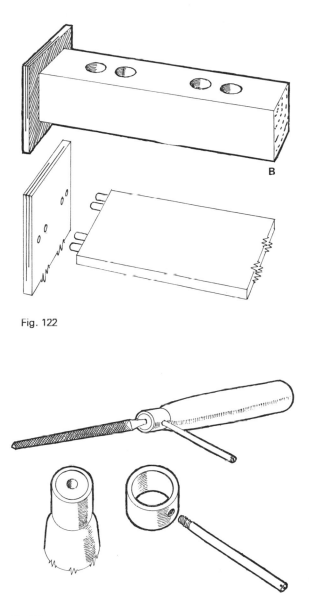

B

Fig. 122

Fig. 123

124 Wood screw gauge

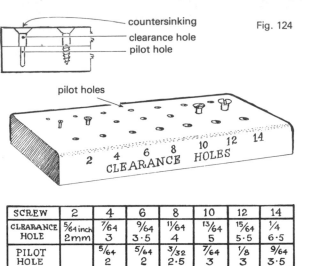

Fig. 124

The principle of using a screw to join two pieces of wood is very simple but often wrongly carried out. The hole in the upper piece of wood known as the clearance hole should be as small as possible, yet the screw should be free to turn in it by finger grip. Too small a hole adds friction, making the screwing operation harder, without any gain in efficiency. The hole in the lower piece has a diameter which is that of the core of the screw. This is necessary because unlike a drill, a screw is not capable of removing wood. The core then fits into this pilot hole and the threads bite into the sides. The clearance hole may or may not be countersunk.

This simple device enables the correct clearance or pilot drill to be selected in an instant from either a millimetre or a fractional inch drill set, obviating the use of tables and reference books. A hardwood block, beech for example, is prepared to size and three rows of holes are drilled in it. Two rows are pilot holes and one is clearance holes. Screw in short specimen screws into the centre row so that all the heads stand up about 6mm ($\frac{1}{4}$in). A range from 14 to 2 gauge is generally enough. Brass-headed screws look better and resist corrosion. Number the screws with a felt-tipped marker.

In use simply identify the screw's gauge if not known by inverting head upon head and select drills to fit the corresponding holes in the block. Finally select a screwdriver whose blade width fits the screw slot exactly.

SCREW	2	4	6	8	10	12	14
CLEARANCE HOLE	$\frac{5}{64}$ inch 2mm	$\frac{7}{64}$ 3	$\frac{9}{64}$ 3·5	$\frac{11}{64}$ 4	$\frac{13}{64}$ 5	$\frac{15}{64}$ 5·5	$\frac{1}{4}$ 6·5
PILOT HOLE		$\frac{5}{64}$ 2	$\frac{5}{64}$ 2	$\frac{3}{32}$ 2·5	$\frac{7}{64}$ 3	$\frac{1}{8}$ 3	$\frac{9}{64}$ 3·5

125 A gauge for machine screws

The woodworker may from time to time become involved with machine screws and the tapping of holes for them. A metal gauge similar to Fig. 125 can be made for any type of thread likely to be used. The clearance holes and tapping holes must be found from reference tables and there is an added advantage that either a metric or an imperial size drill may be used as long as it fits or nearly fits the hole.

Mild steel 4mm ($\frac{3}{16}$in) thick is very suitable. Mark out accurately and tap the holes in the centre row with care. This row is useful to identify an unknown thread.

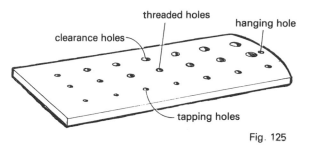

Fig. 125

126 The distinguishing of drills

Fig. 126

Most woodworkers own both metric and imperial drill bits and after some use the stamped sizes tend to become obliterated causing confusion. Imperial drills can easily by distinguished by grinding a small facet on each as shown and as most drills now are black coated this is instantly seen and the drill identified.

127 Awls

Fig. 127

Of this large family of tools from the days of handwork, only the bradawl remains in the catalogues. The convenient materials for making awls are tool steel, silver steel (commonly stocked in good toolshops) or old or unwanted screwdrivers.

The bradawl, **A**, is most used for screw holes and is either filed or ground on both sides and after hardening and tempering (see page 204) is honed to chisel sharpness on the oilstone.

Marking awls, **B**, are made out of thinner material ground or filed to a long, fine and round point. Small electrical screwdrivers with plastic handles convert easily. This awl is not really suitable for work other than marking as it cannot remove wood.

The four square awl or small hand reamer, **C**, is useful in the bigger sizes for enlarging holes and in the smaller sizes for making pilot holes for small screws. It is filed really square in section and after hardening and tempering is carefully honed to give four keen cutting edges. It can be made from either round or square material.

The hooked awl, **D**, is particularly useful for marking out the second stage in dovetails. A good material for making these

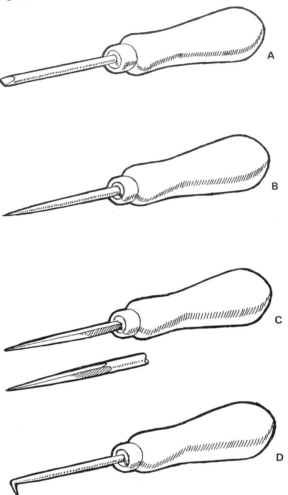

102

is old-fashioned steel knitting needles.

Turned hardwood handles with ferrules are well worth the trouble taken. Fit the blades into the handles by drilling slightly under size, filing the end of the awl to a chisel shape and then driving on in a vice. If the section is big enough the handle and blade can be drilled through and pinned.

128 Tools for producing circular frames

These are commonly built up of three layers, each of six sections and this necessitates the preparation of eighteen identical sections. From a full-size drawing of the frame prepare sufficient strip to width and thickness.

A mitre block with angles of 60° is constructed as at **A**. Pin on a hardboard strip and saw the first mitre. Fix a 60° stop block to give the length of section required then saw out the eighteen sections. Use a template to mark the circumference which can then be sawn out using a coping saw. If a bandsaw or jigsaw is to be used, a pivoting jig, **B**, can be made from the full-size drawing and this can be cramped to the saw table. Two small clipped off pins in the jig will prevent the section from moving and a small clamp holds the work in position.

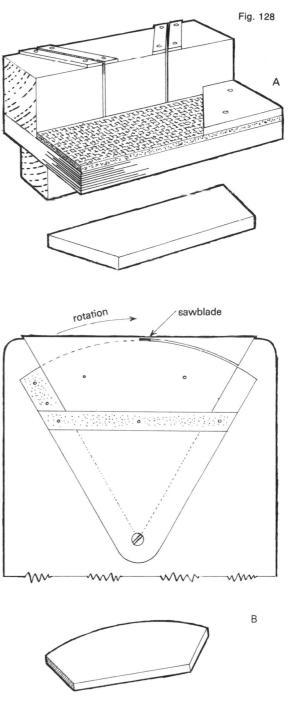

Fig. 128

A

rotation   sawblade

B

A simple scratch tool, **C**, will cut a shallow groove round the curve for the cramping cord, **D**. The sections are pulled together with a simple cramp using nylon, wire or woven cord. The actual tensioner, **P**, is quite straightforward to make and has other applications also. In cramping up the six sections are held firmly between two discs made from plywood, one smaller than the other, which have been faced with adhesive plastic film to prevent any glue from sticking, **F**. They are centrally drilled and attached to a baseblock. The joints are numbered or lettered and then checked and corrections are made on a disc sander or on a sanding disc in a circular sawbench using a 60° guide fence, **E**. This angle can be set up by means of a good draughtsman's 60° set square between the fence and the disc.

When all the joints fit well they are glued up. The baseblock is held in the vice while the sections are glued and assembled. The top disc is fitted in place and screwed tight, **F**. The wire or cord is then tensioned and the top disc slightly slackened to permit the joints to close properly. Six small handscrews as described on page 71, with jaws covered with adhesive film hold down the joints, **G**, and the top disc is tightened down.

Repeat the process for the other

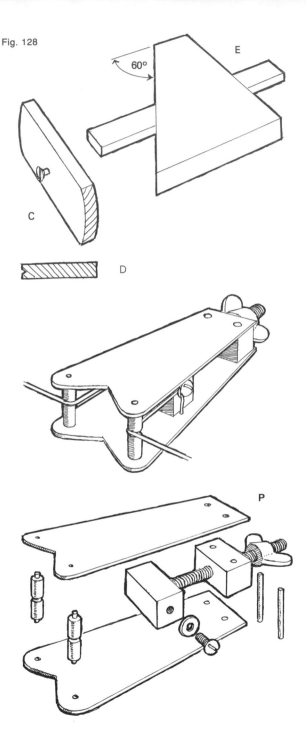

Fig. 128

two layers. At this stage they are very fragile and must be treated with great care. Little cleaning up will be necessary but a jig, **H**, will hold these fragile rings for scraping. A plywood base is gripped by a block beneath it, in the bench vice. A hardwood bar with wing nut and bolt holds the ring firm while a very sharp scraper or block plane levels off any slight irregularities.

The three rings are now glued together with staggered joints cramped between two further chipboard discs covered with adhesive film. To prevent the rings from sliding on the glue they should be held together with adhesive tape before cramping between the discs. As many small G cramps as possible should be used or else a set of special handscrews should be made.

This laminated ring is immensely strong and is now mounted on a large blockboard disc, centred as accurately as possible, with six paper joints. Do not use synthetic resin glues, but rather scotch glue, as this can easily be removed later with warm water. Two rebates are turned, for the mirror and for the plywood back. At this stage the mirror rebate should be painted with a matt black paint. The outer edge and back face are also turned at this fixing, **K**.

The reversing of the ring is not as

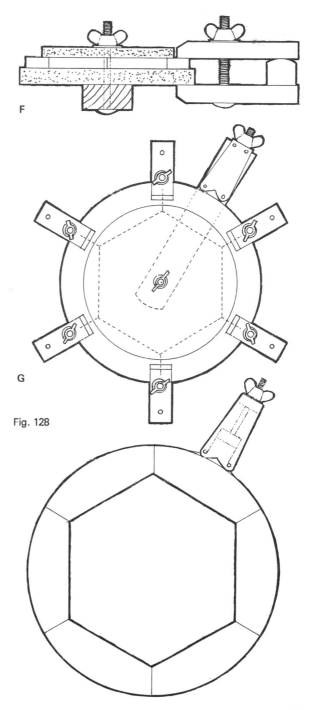

F

G

Fig. 128

difficult as may be imagined. Before breaking off the ring fix three or more locating blocks, **L**. Make these a very tight fit then break the joint. Reverse the ring and either glue again with a paper joint or screw from the back into the centre section. Now turn the front shape and the corner, blending into the sides. The inner edge is trued and slightly angled, also removing any splashes of black paint. Any inlay is put in at this time. For this a special tool needs to be made which can be ground from a light 6mm ($\frac{1}{4}$in) chisel, **M**.

The back, a plywood disc, can be made an accurate fit without a lathe. Six screws secure it, avoid screwing into the joints. The mirror edge should also be blacked. A hanging plate from thin brass can be made as illustrated, **N**. This can screw on, the screws being clipped off and riveted over on the inside. Dust can be kept out of the hanging hole by sticking together a larger and smaller piece of adhesive plastic tape, **O**. The space between the mirror and the back can be packed firmly with small pieces of plastic foam.

Details of the construction of a wire or string cramp are shown at **P** and the method of use at **G**.

Fig.128

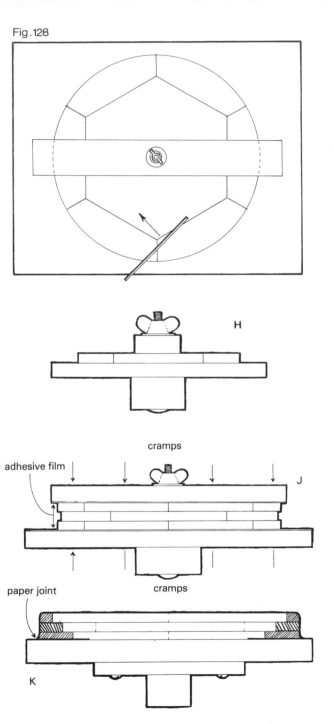

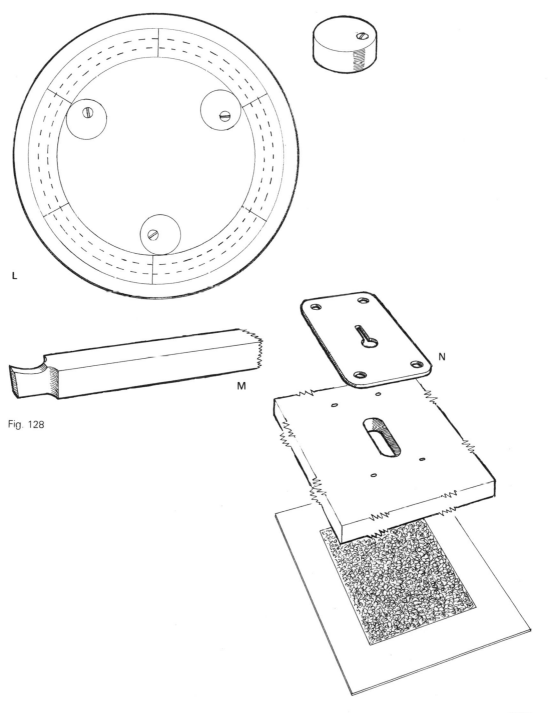

L

M

Fig. 128

N

129 Mullet

When a frame is being grooved for a panel, either by plough, router, circular saw or spindle moulder, it is advisable to groove an offcut of hardwood with the same setting. This is a mullet. When the panels are being fielded this is used to test the edge thickness. It can be easily slid the length of the panel and is more convenient than using a frame member.

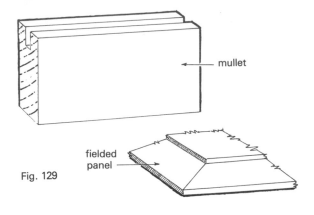

Fig. 129

130 A scribing knife

The scribing of a line across a cavity or recess is a task for which no suitable tool can be bought. One of the ways the knife can be used is shown at **A** but there are several others. It is particularly useful for cutting joints in veneer sheets using a hardwood straight-edge.

The leaf-shaped blade can be made from tool steel which is marketed in lengths of 20 × 2 or 3mm ($\frac{3}{4} \times \frac{1}{16}$in) or from a machine hacksaw blade. The latter must not be a high-speed steel blade. Soften by heating to bright red then allow to cool slowly. In this state it can be slowly sawn, drilled, countersunk then filed to shape. A block of the hardest available hardwood is used for the handle. This is prepared marginally wider than the blade and has a recess cut marginally greater than the blade thickness.

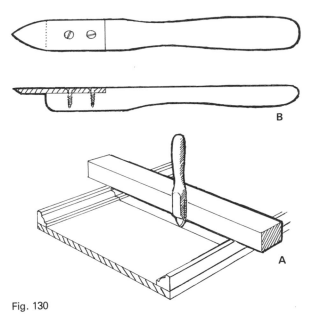

Fig. 130

108

The blade is now hardened and tempered as described on page 204. Screw the blade and handle together with countersunk steel screws. The head of the screw must stand proud so that when filed flush the slot is removed. The points should come through and are riveted slightly. The handle is now filed or glass-papered on a wood block until blade and handle are truly flat, **B**. The remaining sides and end of the handle are shaped to choice. Finally the blade is ground and sharpened, still keeping the flat side true, and the wood finished either with oil or polyurethane varnish.

131 The rip backsaw

A board of a thickness of about 75mm (3in) is being sawn along the grain at **A**. The saw shown is a ripsaw of $4\frac{1}{2}$ teeth per inch which is the correct saw for the job as its teeth are sharpened specifically for this kind of cutting. A tenon with a width the same as the board thickness in **A** is shown at **B**. This is normally sawn with a tenon saw of about 14 teeth per inch, these teeth being cut specifically for cutting the other way, i.e. cross-cutting. This makes the cutting laboriously slow and the longer the saw is worked in the kerf the wider it becomes and the more likely is the saw to wander from the line. The answer is to have a

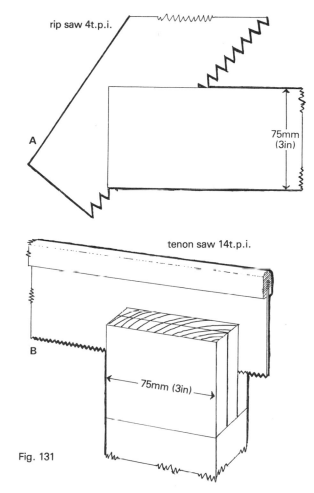

rip saw 4t.p.i.

A

75mm (3in)

B

tenon saw 14t.p.i.

75mm (3in)

Fig. 131

matched pair of tenon saws, one cut as a cross-cut of 14 teeth per inch and the other recut, see Fig. 192, to 10 teeth per inch and filed straight across giving ripsaw teeth. Recut in this way tenon cheeks are cut quickly, easily and cleanly. For ease of identification the handles of rip tenon saws can be dyed black, while cross-cutting tenon saws retain the natural unstained handle.

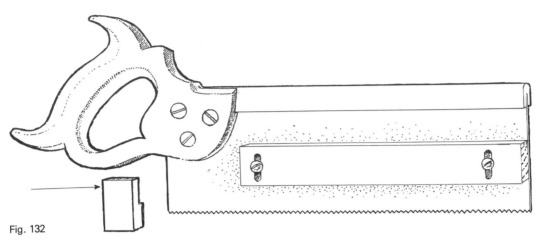

Fig. 132

132 Backsaw depth stop for housings

A 300mm (12in) backsaw is most suitable and preferably an old one. A hole is drilled near each end with a very sharp drill. Hold the blade between two pieces of mild steel or hardwood to prevent the blade distorting during the drilling. A slotted fence of hardwood can be fixed either with saw screws or with screws and wingnuts. Set the exact depth by means of a hardwood block as shown. The method of use is obvious.

133 Veneer strip cutter

As with the planes described on page 84 a built-up construction is used, **A**. The mouth or gap is prepared square before glue-up and it is most convenient to make the gap the width of 14 point printer's space blocks. Al-

ternatively wood spacing can be used with veneer slips for fine adjustment. One piece of spacing should be angled to match a wedge of rosewood or something similar, the arrangement is shown in section at **B**. The cutters are the blades for cutting gauges. If these are not available, cutters can be sawn and filed from cabinet scrapers.

The veneer to be cut is held on a prepared cutting board of chip or blockboard by a strong, straight-edged batten. The tool runs along this as in **C** to cut the required strips. Sizes can suit personal taste and available material. If convenient, two ready-made handles can be fitted. If the centre section is made 50mm (2in) in width then strips for veneered chess board can be cut with ease. Cross-banding can also be produced easily.

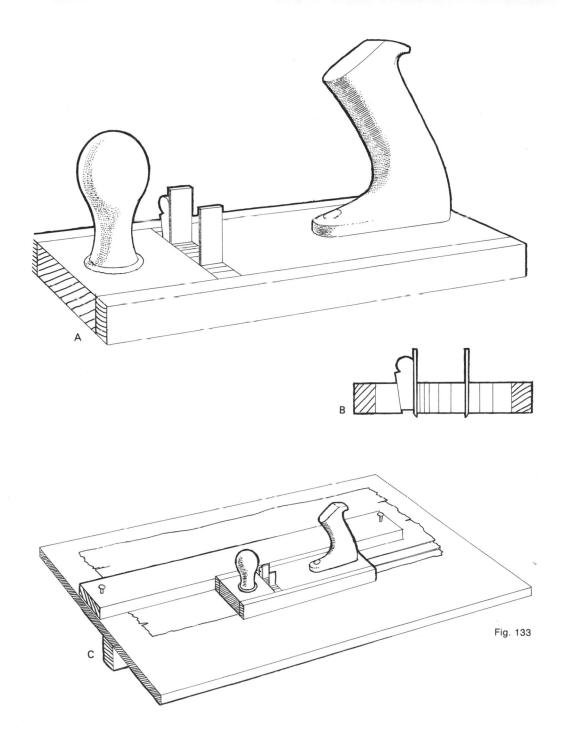

A

B

C

Fig. 133

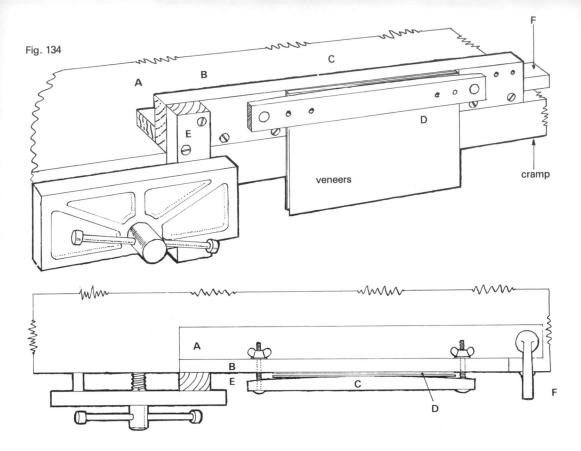

Fig. 134

veneers

cramp

134 Veneer jointing – a planing aid

Of the various methods of jointing veneers, planing is the most successful. Veneers ironed on with a glue film or applied with a vacuum bag press, must be jointed and taped together first.

The base, **A**, has an upright block, **B**, secured to it. One end, **F**, overhangs to permit cramping to the bench top. Block **E**, screwed to **B** at the other end is held securely in the vice. The clamp bar, **C**, is planed to a definite curve on its inner surface. Several holes in the clamp bar match up with corresponding holes in the block **B**. Bolts and wing nuts, as close to the ends of the veneers as possible, tighten the jaws to grip them firmly. The veneers should protrude 1–2mm ($\frac{1}{16} - \frac{1}{8}$in) and are best planed true with a sharp trying plane until the slightest shaving has been removed from the jaws.

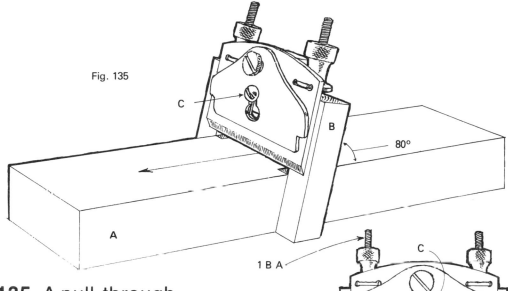

Fig. 135

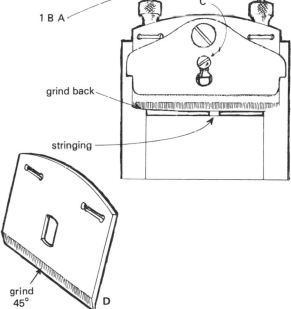

135 A pull-through thicknessing scraper

In some ways this tool can be considered an adjunct to the device for producing stringings on the sawbench, Fig. 177, page 185. It can be used on these or hand-sawn stringings and strips. It was originally a guitar maker's tool used to produce very fine strips from which to build up decorative end-grain mosaics.

The base, a stout hardwood block **A**, is produced slightly narrower than the spokeshave blade which will be used. The bridge piece, **B**, is built up and housed and glued into the base at an angle of 80°. A round-headed wood screw holds both blade and cap in place, **C**. Thumb nuts and screws are fitted as shown. These may be obtained from an old spokeshave, in which case the thread will be 1.BA. One corner of the blade, as indicated, is ground back slightly for ease of entry. After sharpening keenly, the edge is burnished in the manner of the scraper plane, **D**.

The stringings are entered on the left and moved to the right when pulled through. To avoid a thick end, pull each piece through twice starting at opposite ends. It is easier, with the very fine adjustment available, to fit the stringing to the groove cut by the scratch tool or turning tool than it is to make the tool to suit a particular piece or batch of stringing.

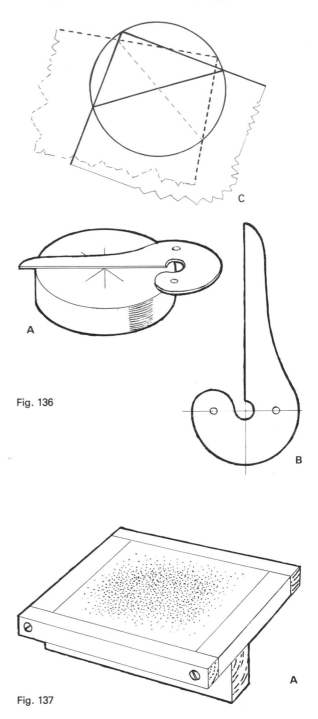

136 Centre finding

The common pattern of centre square is shown at **A**. It can be quite small, as the engineer's steel centre square. A larger one can be made from plywood and in this form is most useful for finding the centres of sawn discs for woodturning. The centres of the holes are on a line at right-angles to the marking edge **B**. A second method is shown at **C**. A true right-angle is prepared on a piece of card or thin plywood. The right-angle corner is held on to the circumference and marks made where the two arms cross the circumference. These join to give a diameter. Repeating will give an intersection at the centre.

Fig. 136

137 Glasspapering aids

The glasspaper board, **A**, is useful for such jobs as flattening small items and trueing up bases of turned objects. The base is of 19mm ($\frac{3}{4}$in) blockboard with the

Fig. 137

usual vice grip beneath it. Glass-paper is tucked under the two screwed-on clamp strips and is easier to change than if glued on and no damage can be done as could be the case if staples were used.

The square-edger, **B**, is used to sand thin edges, both curved or straight, when it is very easy to lose this squareness. A coarse and a fine strip of glasspaper are held in place on each upright which are screwed to a base made from 19mm ($\frac{3}{4}$in) block-board. The upright not in use is held by the vice leaving two hands to control the work. This device can also be used after using a disc sander to remove the circular scratches.

The sanding plane, **C**, requires no great explanation. Standard knobs on a base of multi-ply or blockboard comprise of the top. If the base is covered with an offcut of plastic laminate and the glue used is a scotch glue, the worn paper can be easily soaked off and replaced.

Fig. 137

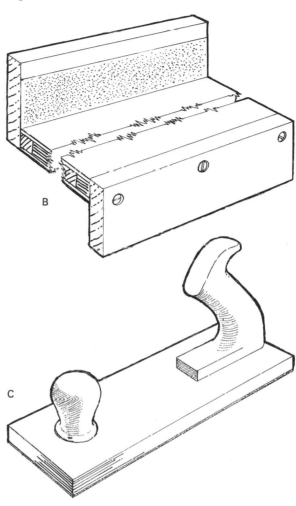

138 Glasspaper shooting board

This device was developed initially to true up the mitre joints of smallish picture frames. However well-made and maintained, mitre blocks cannot be relied upon always to produce perfectly-fitting joints. It is not intended that the glasspaper plane should take off more than the smallest amount of material and this tool is certainly preferable to the plane when working on highly decorative mouldings built up with plaster.

The board itself, **A**, has a base of blockboard with a stout block fixed to its lower face in order to hold the device in the bench vice. The next layer is of 8mm ($\frac{3}{8}$in) plywood and above this is the top layer or working surface of 8mm ($\frac{3}{8}$in) or 13mm ($\frac{1}{2}$in) plywood. The stop block of, say, 20mm ($\frac{3}{4}$in) is fitted centrally giving angles of 45°. More important than the accuracy of the two 45°s is the fact that the total of the two angles must be an exact 90°.

The plane itself, shown at **B** and **C** is built on a base of 8mm ($\frac{3}{8}$in). A stout vertical block of say 25mm (1in) hardwood is glued and screwed to this and it must make a true right-angle. This block is faced with a plastic laminate strip and it is probably better to fix this first, then check the angle, before gluing. Make sure

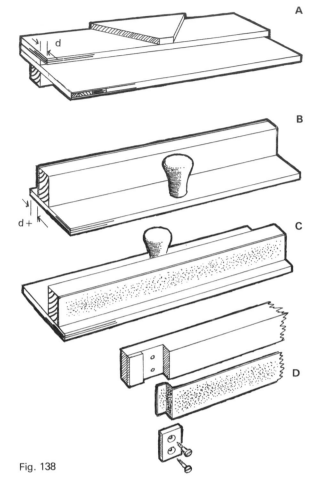

Fig. 138

that the lip of the plane, d+ is slightly greater than the groove, d, into which it will fit. The plane sole may need a very gentle sanding in order to slide sweetly in the groove. A round handle permits equally convenient left- or right-hand working.

Glasspaper can now be glued to the laminate face using a scotch or other animal glue in order that well-worn glasspaper may be soaked off. Heavy scoring of the laminate face will give better adhesion. The glasspaper chosen should not be of too fine a grade. A 60 grit is a compromise between too course a cut and too short a life. To avoid gluing the paper an improvement is shown at **D**. A small brass or steel cramping plate is screwed into a housing which is slightly deeper than the combined thicknesses of the metal and the glasspaper. A small strip of veneer, or even the glasspaper itself will level up the cramp.

Other applications of this tool will occur using stop blocks of 90° or any other angle. If more than just a trim is required the conventional planing will be more suitable since the glasspaper will soon begin to wear over a narrow area.

139 Saw guards

Saw sharpening is a tedious job and much of it can be avoided by taking more care of the saw and

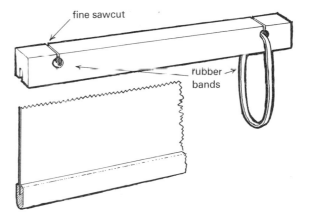

Fig. 139

avoiding contact with the teeth on metal objects or tools. A simple guard as illustrated does this. Having prepared a hardwood strip, gauge it heavily down the centre of one of the narrow edges with a marking gauge and run the saw down this groove to a depth somewhat deeper than the teeth. Drill two small holes as shown and make a thin sawcut into each. A tightly-stretched rubber band can be pulled into each and these secure the guard far better than loops of string and can easily be replaced.

LATHE
AIDS

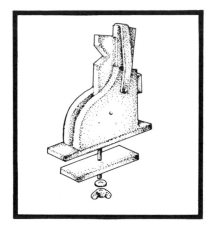

140 Dowelling, boring & sanding on the lathe

All these operations can be done easily and accurately on the woodturning lathe and they all require a table, adjustable in height, except sanding which can be satisfied with a fixed table of disc centre height. The base, **A**, is a working fit in the lathe bed. A round bed lathe will require **A** to be permanently fixed to the table with the sides screwed off and on as required as in section **R**. The fixed sides **B** are glued and pinned to this, but before fixing they and the moving sides **C** have matching shallow grooves cut as shown. A hardwood strip glued into **C** enables the top section to slide up and down **B**, to give height adjustment which is fixed by bolts and wing nuts. A strip of cartridge paper placed at this stage between **B** and **C** will make for an easy working fit later. The blockboard top **D** is glued and pinned to the sides **C** and can, if necessary, be faced with a plastic laminate. A rebated block, **E**, with a long bolt and wing nut secures the table in place.

Four fences are shown but others can be devised for special jobs. A straightforward dowelling or drilling fence, **F**. With a drill in a drill chuck in the lathe headstock and the table set to the right

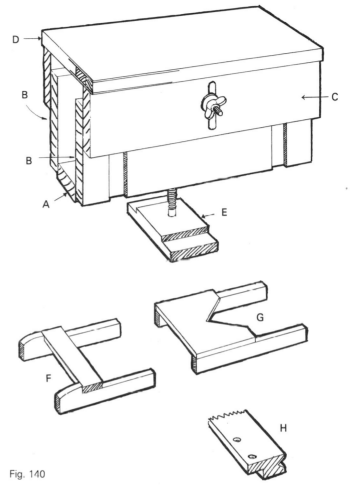

Fig. 140

height, **F** will feed up the workpiece at right-angles to the drill, for work such as **f**. Fence **G** will feed discs such as turned lamp bases for drilling radially as in **g**. **H** can be screwed to **F** and set accurately to the drill centre height, to drill spindles as at **h**. Fence **J** can be built up, and is secured by thumb screws with big washers to one of several pairs of tapped holes in the table top. Position this and line up accurately parallel to the drill axis which could be marked on the table top. This enables end-grain components to be dowelled as **j**, or sanded.

A sanding disc can be made by screwing a plywood or blockboard disc to one of the inside faceplates. Turn true then face with a plastic laminate. Turn the edge true then score concentric circles on it which help with the gluing on of the disc. Sanding discs without a centre hole can either be obtained commercially or cut out from the sheet. Glue the discs on with a scotch or animal glue. This allows worn out papers to be soaked off and the disc cleaned which could not be done using a synthetic resin glue.

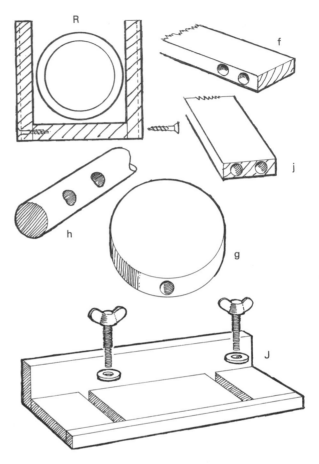

Fig. 140

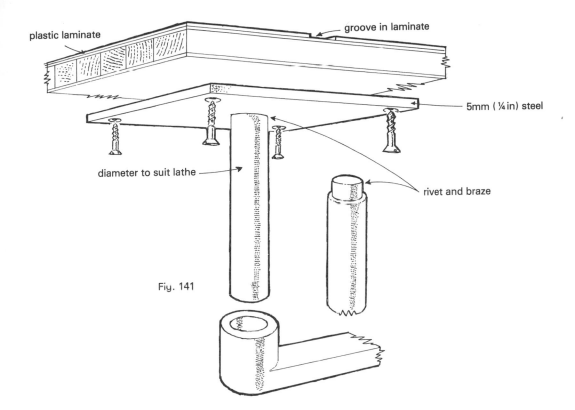

plastic laminate

groove in laminate

5mm (¼in) steel

diameter to suit lathe

rivet and braze

Fig. 141

141 Simple sanding and boring table for the lathe

A steel column must be found which fits the lathe's toolrest clamp. To this is fitted a square steel plate of say 100 × 100 × 5mm (4 × 4 × ¼in) which is riveted or brazed or both. To this assembly is fitted the blockboard table faced with a plastic laminate. A groove, as on a saw-bench, may be provided and if so the table should be carefully lined up with the sanding disc.

142 Lathe chuck guard

Small wooden or metal items often need to be polished in the lathe. These can easily and quickly be held in the three jaw chuck. However, holding the glasspaper or emery cloth in the hand is dangerous as nothing is easier than to knock the knuckles on the chuck jaws. This guard, **A**, makes it possible to sand and polish even the smallest knob in complete safety.

Make the base plate from blockboard or 20mm (¾in) hardwood. Glue and pin below this a suitable block to fit closely into the lathe bed. The upright guard plate is from 8mm ($\frac{5}{16}$in) plywood. A small drill in the chuck will give the centre on which to bore the hole, say 25mm (1in) in diameter. The guard is shown in position at **B** with a small knob in the chuck as the workpiece. The base must be cut to such a length that when it is pushed hard against the lathe headstock the chuck is just cleared. A long bolt through the lathe bed, with a wooden block could be used to lock the guard firmly but in practice this has not really been found essential.

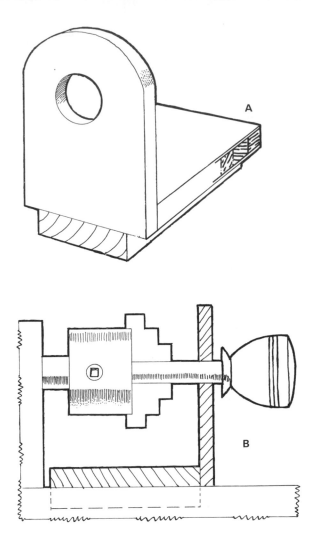

Fig. 142

143 Lathe steadies

The turning of long thin components is made difficult and sometimes impossible by whip but this can be overcome by supporting at the middle by a steady. **A** and **B** show a bolt-on model for a round bed lathe. The pivoted notched support is kept in contact with the work, in theory, by a wedge behind it. Two small woodscrews holding a rubber band over the top keeps the support firmly against the work. A similar arrangement, **C**, serves for lathes with a conventional flat bed. An alternative method using two roller bearings shown at **D** make either an improvement or an unnecessary complication, according to preference.

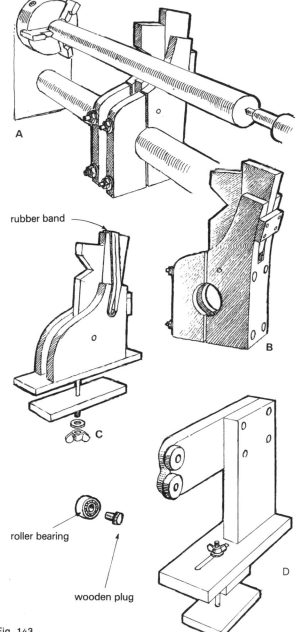

Fig. 143

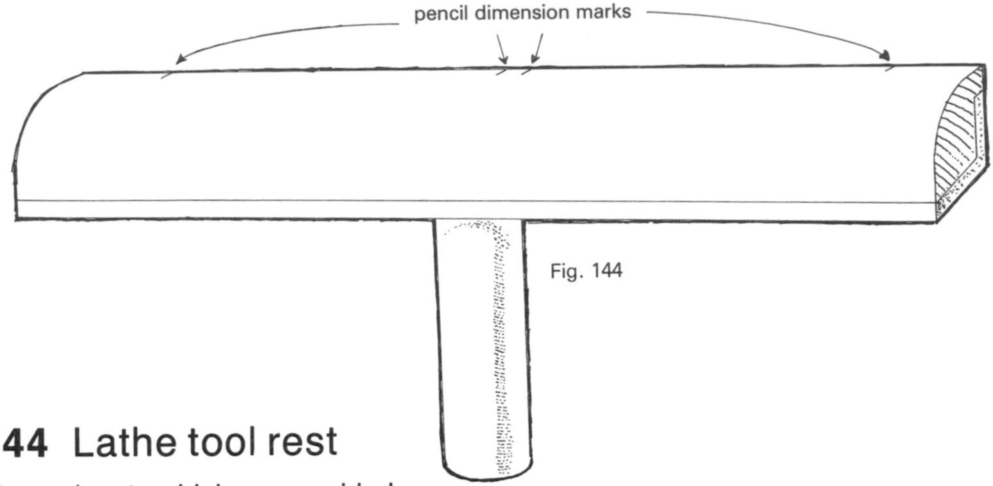

pencil dimension marks

Fig. 144

144 Lathe tool rest

The tool rests which are provided with some lathes are far from perfect, often giving the tool two distinct fulcra. The model shown is of bright angle iron to which is welded an upright of a suitable diameter. The wooden fill-in is firmly screwed to the iron. It is then shaped to a smooth curve of approximately a quarter circle. Quite a length can be made without losing stability. The tool, in particular the gouge, slides very smoothly on the wood surface and if it gets nicked it is very easily trued up with a plane. For repetitive work pencil dimension marks are quite a help.

Lathe chisels when new have very sharp corners and these will catch in any slight unevenness on a wood or metal tool rest. The two leading edges of the chisel should, therefore, be ground and stoned to a smooth curve and the chisel will then slide as sweetly as a gouge.

145 Long hole boring in the lathe

There are a number of long hole boring devices on the market. Most of these are both unnecessarily complicated and unnecessarily expensive. The turning of a hollow tailstock centre, as illustrated, is a simple and straightforward job. It is tapped to take a screw, brass so as not to damage the tailstock barrel on which it fits. The centre should be case hardened, see page 204. On larger lathes a morse taper tailstock can be similarly bored

Extended drill The hole is generally made with long shell auger but this is not essential. An ordinary twist drill, 8mm ($\frac{5}{16}$in) is the size commonly sold for table and standard lamps, can be extended on the lathe. The drill end is turned down to about half its diameter and a length of mild steel rod of the same diameter is bored to take the drill, which can be brazed in. A wooden handle should be fitted.

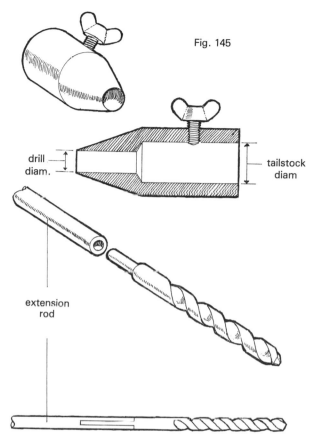

Fig. 145

drill diam.

tailstock diam

extension rod

146 Intermediate faceplates

Bowls and platters which are turned with a completely flat base screwed to the faceplate are only really half-turned. A well-turned bowl, like a pot needs a base ring. For this the outside of the bowl must be turned first then the bowl is reversed while the inside is turned. This requires very accurate re-centring. Normally the bowl is reversed on to a wooden disc and held there with a glued paper joint. This means an overnight wait while the glue sets and a separate disc to be turned for each bowl. Intermediate faceplates can be used immediately and the plates themselves can be used indefinitely.

The first stage is to set up in a simple metal-turning lathe all the standard faceplates and turn out from the centres identical shallow recesses of about 40mm (1½in) by 2mm ($\frac{1}{16}$in) deep, **A**. The intermediate faceplates can be cast from aluminium or iron. The former is well within the capability of most school or college workshops. For these, wooden patterns will be required, **B**, somewhat larger than the faceplates themselves and of extra thickness to turn down to the approximate finished dimensions shown at **C**.

The back is turned flat and the small locating peg made to be a

Fig. 146

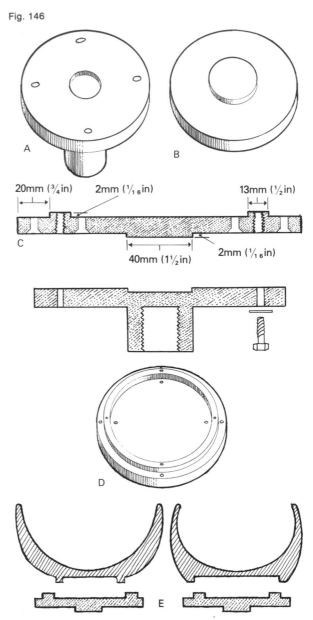

tight fit on the faceplate. Cramp the two together and with a drill that just fits the faceplate screw holes drill a small dimple in the casting. In this dimple suitably drill and tap the casting which is fixed to the faceplate with a hexagonal-headed screw. The other four holes are now similarly dealt with. This pair can now be held in a three-jaw chuck and the face trued up.

On each of the intermediate faceplates a small upstand of about 2mm ($\frac{1}{16}$in) is turned. Each is of a different diameter covering as wide a range as possible. Drill eight clearance holes for No. 10 screws, four about 6mm ($\frac{1}{4}$in) outside and four 6mm ($\frac{1}{4}$in) inside the upstand. Countersink deeply on the faceplate side, see D.

To use, turn the outside of the bowl in the normal way on the standard faceplate. The small central cavity has no effect. Now turn the base ring to either pattern shown at E to be a tight fit on the intermediate plate most near to the desired diameter. If necessary, polish at this stage then screw on the intermediate faceplate. This is now screwed to its standard faceplate and the turning of the inside can start at once. On completion make four small plugs with a plug cutter either of the same or a contrasting wood. Drill a shallow hole for these, glue in and finally clean off.

If, as is often the case, the four screw holes in the basic faceplate are not truly spaced, time can be saved by filing a register mark on the rim of the standard and the auxiliary faceplate.

147 Brush boring jig

Brushmaking is an activity to which woodworkers sometimes apply themselves. The basic problem of brushmaking is the boring of the holes. A very simple brush can be bored vertically on a drilling machine but the bristles as shown at B do not look particularly elegant. A better looking brush as at C results when the holes are so drilled as to splay the bristles. In this case only the central hole is truly vertical. The device for arranging this, A, can be quite easily made to suit the average woodturning lathe. The end is anchored into the tailstock either by means of a long bolt, E, or by an old morse taper drill which has been softened by heating and sawn off, D. On to this anchorage is fixed a steel tube, the end of which has been filled. It can pivot on two screw eyes as in F or on a two-way pivot which naturally requires more work and machinery, G. The end of the tube has a small screw eye from which the device is later suspended. The tube can be about 300mm (12in) long and into this slides a steel rod, 10mm ($\frac{3}{8}$in) is a convenient size. On to the end of the rod is brazed a

Fig. 147

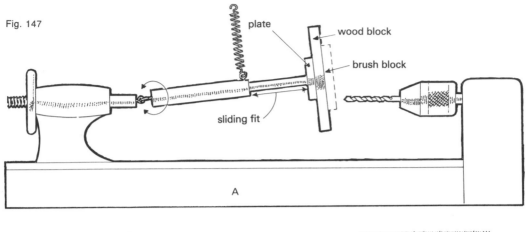

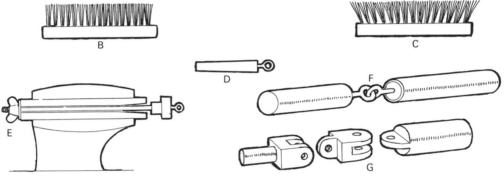

steel plate of about 150 × 75 × 3mm (6 × 3 × $\frac{1}{8}$in) and to this is screwed a hardwood block somewhat larger than the brush being drilled. This needs to be carefully centred and on to it is screwed the marked-out brush block. If this is made overlong it can be screwed through the waste ends. The assembly can be suspended on a cord with a weak coil spring.

A good twist drill can either drill right through into a piece of waste wood or drill part way with a wooden depth stop. To operate, stand behind the headstock and pull the workpiece on to the drill.

Some form of guard is needed to prevent garments being damaged by the revolving end of the headstock spindle when leaning over the headstock.

DRILLING
AIDS

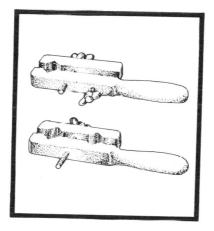

148 Aid for drilling corner holes

It is a common problem to have to drill holes in the corners of a rectangle. Cover plates, name plates and the like are examples. The jig consists of a strong well-jointed right-angle of hardwood, possibly from 25 × 20mm (1 × ¾in) material, screwed from below to a baseboard of chipboard or plywood, large enough to conveniently cramp to the table of a drilling machine. From the inside corner is marked a 45° line. All holes bored will have their centres on this line. An indicator is needed and can be turned to a sharp point from a piece of mild steel, say 8mm ($\frac{5}{16}$in) diameter. Put the indicator in the chuck and place the jig in the required position, checking that the indicator is on the 45° line. Cramp firmly in this position. Put a waste piece into the corner and adjust the depth stop. Now drill the corners successively, into the waste piece.

149 Drilling vices

The woodworker does not use the drilling vice to anything like the extent of the metalworker. Nevertheless these wooden vices are invaluable for holding very small pieces on the drilling machine. Model **A** is a development of the handscrew and holds a single small piece. Model **B** is simpler and holds two small com-

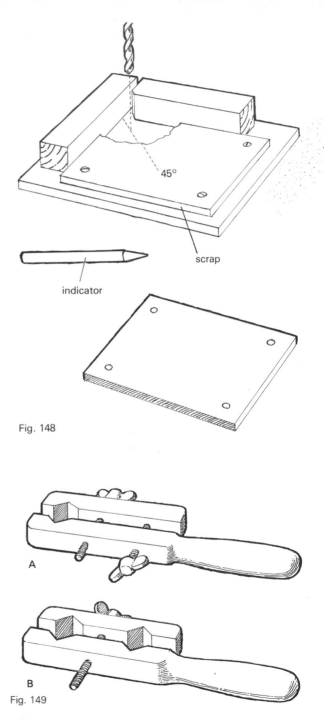

45°

scrap

indicator

Fig. 148

A

B

Fig. 149

ponents or one with some packing and model **C** holds flat members. If the jaws are made of good hardwood the screws cut into them will last for years.

150 Vertical drilling aid for the drilling machine

Machines having a table which swings horizontally make it impossible to drill on to a centre fitted to the table, so to ensure truly vertical drilling an aid must be used. A vertical and a horizontal piece of blockboard are glued and pinned together, giving an exact right-angle. The angle is made secure by pinning on two plywood webs. A further block is fixed truly vertical, at right-angles to the base. Work cramped tightly into the corner so formed will be drilled true. The work can be reversed and a long, series twist drill used for greater depths. Sizes must be chosen to suit the machine, available timber and type of work envisaged.

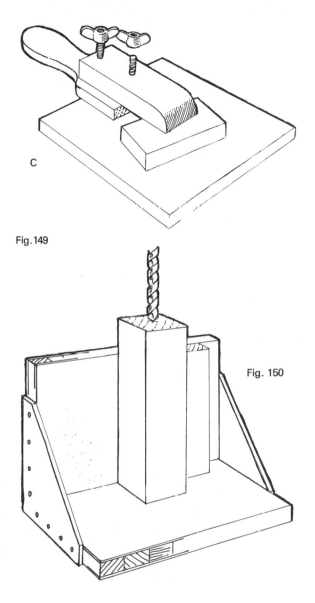

C

Fig. 149

Fig. 150

151 Drilling aid for vertical cylinders

This is similar to the normal vertical drilling aid but with the difference that a vertical cradle is needed to hold round work. The job itself is held by a stirrup cramp which is described in Fig. 153. A shaped block is needed to prevent damage to the work.

152 Diametric drilling of spindles

Hand drilling or boring through round spindles is a chancy business. Perfection is achieved by putting through a small pilot hole first. This goes through halfway from each side. The starting points, exactly diametrically opposite can easily be located by wrapping round a strip of paper. The circumference is marked, then the strip is folded in two and so marked. The paper strip is replaced squarely and the two opposing starting points are marked.

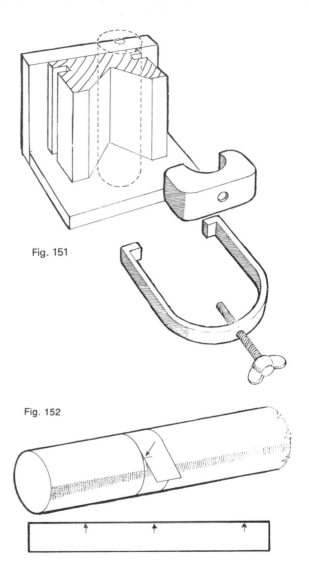

Fig. 151

Fig. 152

153 A drilling cradle for the drilling machine

Like all cradles this is most easily made from two sections. The sides are grooved to take a stirrup cramp which can be forged or bent from any suitable material, say 20 × 4mm ($\frac{3}{4}$ × $\frac{3}{16}$in), with a nut brazed on top. A wing nut is brazed to a suitable screw to hold the cramping blocks. Cramping blocks of various sizes are made by boring small pieces of hardwood, sawing in two, shaping then drilling a small cavity in the top. The plywood baseplate to the cradle is needed to give clearance between the fixed block and the stirrup cramp.

For accurate work, particularly for drilling a number of holes in line, an indicator of mild steel is held in the chuck and the cradle centred on its point. A slotted block, secured to the drill table, permits the cradle to be slid with its workpiece back and forth yet always central below the drill. In use check that no chips from the drilling get in between the cradle and its guide block at **x**.

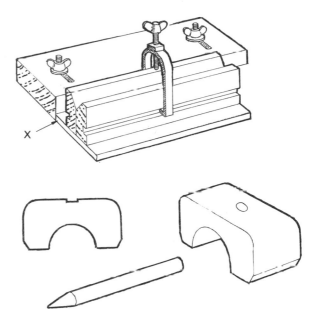

Fig. 153

133

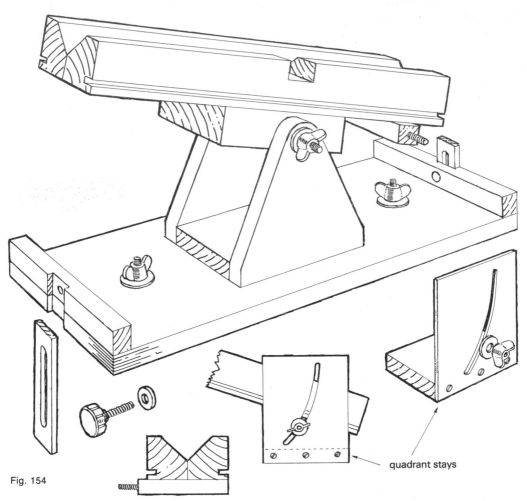

Fig. 154

quadrant stays

154 Tilting cradle

Chairs, stools and tables with splayed round legs present the problem of drilling several accurate angled holes for the stretcher rails. A tilting cradle is the answer to this. A baseboard of plywood or chipboard is made slightly longer than the table of the drilling machine. Depending on the shape and style of the table a means must be found to hold this board firm. A small hardwood block fitted to each end is housed to take the two slotted support arms. These blocks are later threaded to take the locking screw. A convenient size would be 10mm ($\frac{3}{8}$in). The cradle itself is similar to that described previously and can be made in two parts. The sides are grooved to take the stirrup cramp, see Fig. 153. A block glued centrally underneath takes the pivot bolt.

The trunions are best made from thick plywood and are glued and screwed to a block of the same material as the pivot block. It makes sense to assemble the trunions and the pivot block and drill by machine through them all together. Thus all the holes are in line. A long 10mm ($\frac{3}{8}$in) coach bolt with wing nut and washer completes this part.

Fix the baseboard to the drilling machine table and line up the cradle exactly below the chuck. This can easily be arranged by holding in the chuck the steel indicator pointer as in Fig. 153. Glue and screw in this position. It is important that any holding down bolts should be a very loose fit in the baseboard holes. This permits the slight movement which may be necessary to reposition the cradle exactly under the indicator. To prevent movement during drilling two slotted support arms are now required. With modern adhesives it is often easier to build up these rather than cut out the slot. These fit the housings previously cut in the end blocks of the baseboard projecting below the machine table. For a small bench-mounted drill a quadrant stay will have to be devised. Good sized knobs can be made from hardwood tapped for a tight fit of the screws which are held in by epoxy resin glue.

There are three ways of adjusting the cradle to the required angle. Small plastic angle indicators (clinometers) are available similar to the very expensive main member of an engineer's combination set. The required angle may be cut as a plywood template. This, interposed between cradle and a small spirit level, will set the angle. Some draughtsman's adjustable set squares will work equally well. If from a drawing the incline is known, say 1 in 10, a plywood template can be similarly made.

To drill a second hole parallel with the first, move the component along the cradle to the desired position. Insert a dowel or steel rod into the first hole and check for vertical with the spirit level. To drill at right-angles to the first hole, again insert the dowel and rotate the workpiece and dowel until the latter is horizontal by level. It may be necessary to cut a notch in the cradle to do this. For other angles adopt a similar procedure, using an angle template between workpiece and level.

Fig. 155

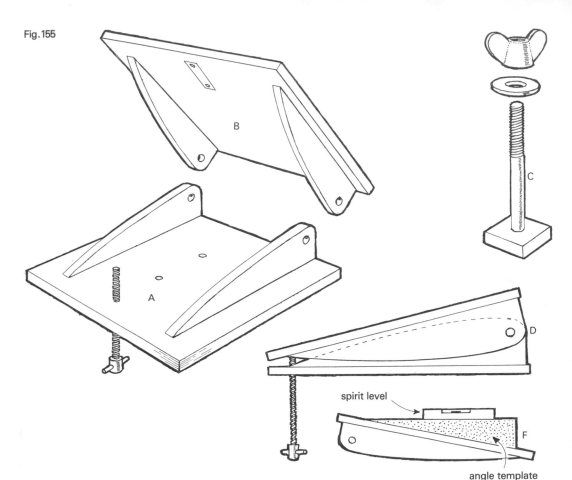

spirit level

angle template

155 Tilting table

The tilting table is the most common method of drilling angled holes in flat components. The base, **A**, is of plywood or blockboard. Bolts as at **C** will secure it to the drill table. Two sloping blocks are glued and screwed on to provide the pivots. The top, **B**, is similar and fits over **A** with bolts through the pivot holes. The base is fitted with a suitable adjusting screw. The top has a metal plate fitted to take

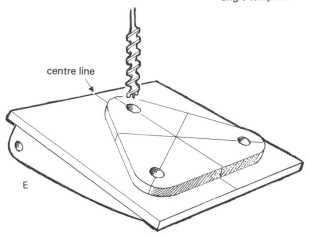

centre line

the wear of the screw. A side view of the assembly is shown at **D**. The top has a centre line marked. **E** shows a stool top with the hole centre lines marked and one lined up for drilling. Angles can be set with a clinometer or angle indicator such as in an engineer's combination set. If these are not available make an angle template and use with a spirit level as in **F**.

156 A drilling table centre

Long holes and slightly angled holes can conveniently be drilled on a machine using a table, **A**, fitted with a centre or spike, **B**. A diameter of 13 to 15mm ($\frac{1}{2}$ to $\frac{5}{8}$in) is suitable for fitting into a central hole in a false table from 20mm ($\frac{3}{4}$in) blockboard.

There are two methods to centre the point. Either grip the centre in the chuck, wind up the table (or wind down the drill head) position the blockboard by its hole, cramp firmly in place, then lower the table to the required height. Or, if the table does not rise truly vertically, fix it at the required height. Then position the centre under the chuck by means of a pointed plumb bob. Hold the workpiece on the centre and drill as in **C**. For angled holes mark with a cross the amount of offset on one side of the workpiece. Position this mark on the centre and then drill as in **D**.

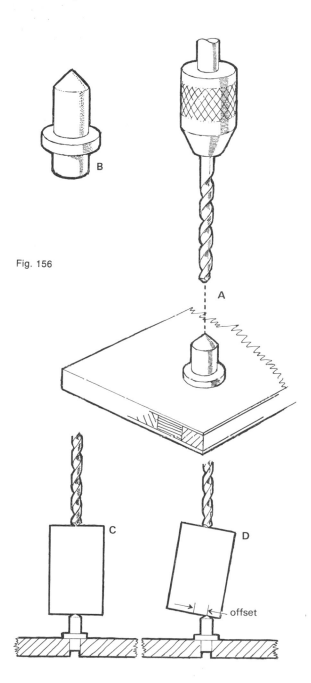

Fig. 156

157 Co-ordinate drilling

This is a method of drilling angled holes which are difficult to accurately measure or apply, by using the displacement from the vertical, applied to a grid in the manner of map co-ordinates, hence the name. Instead however of north/south and east/west co-ordinates, we use left/right and front/back. The principle applied to a very simple solid topped stool is shown at **A**. The ideal bit for angle work is the saw-toothed machine bit. The cost of these is very high and quite a satisfactory alternative is the fast cutting centre bit, **B**. Flat bits are also worth experimenting with; the long point needs to be reduced.

There are two essentials for this system. The first is an extension or guide rod. The second is a guide bush to hold the rod at the correct angle. Three boring tools are shown at **C** but no doubt the reader will be able to devise others to suit equipment available; **a** is a normal brace on to which the guide rod has been welded. The wooden knob is removed, bored then returned. The second, **b** is a simple chuck with a grub screw made from a suitable piece of hexagonal steel. The other end is threaded to take the guide rod which is suitably screwed and given a screwdriver slot. Screw-nosed bits invariably

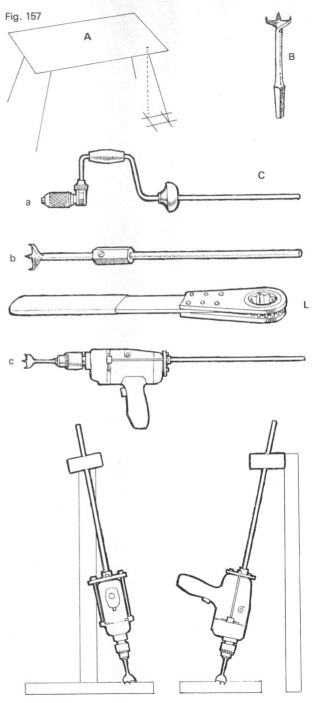

Fig. 157

have a round shank after the square end has been sawn off, so extra chucks can be made to suit bits of differing diameters. This chuck works with a rachet spanner, **L**, and it is advisable to obtain the spanner first then match the chucks to it. The third, **c**, is an electric drill fitted with a guide rod. The adaptor must be made to suit the particular drill. The one illustrated at **D** is for the Black & Docker, with 13mm ($\frac{1}{2}$in) chuck. This chuck will take the saw-toothed machine bits. If, however, the centre bits are used the screw nose must be filed off to give a shortish, square, pyramidal point. The guide bush is in the centre of a gimbal mounted on post of 75×50mm (3×2in) held in the vice. The centre line should be clearly marked and so should a distance of 500mm (20in) from the centre. The arrangement is shown at **G**. Finally a grid is needed of 10mm (or $\frac{1}{2}$in) squares. This can be scratched on a piece of thick polythene sheet, **E**.

Method of use Plot on the grid the hole position, then the displacements left/right and back/front (which are often the same). Fix the grid to the job, parallel to the centre lines. Set the guide bush at the correct height above the job. If this height is insufficient for the drill or brace to fit in then increase the displacements proportionately for a height of 500mm (20in). Put in the plumb

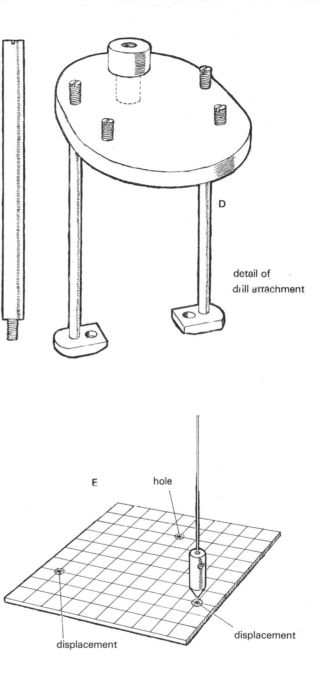

D

detail of
drill attachment

E

hole

displacement

displacement

Fig. 157

139

bob and cramp down the work-piece with the appropriate displacement mark under the plumb bob. Remove plumb bob and grid, fit in the guide rod, and drill on the hole mark. Holes with a considerable angle are best started with the drill free from the gimbal and held vertically. Slowly tilt until the guide rod touches the gimbal. Then thread through and proceed as described.

Repeating aid Plumbing each hole is rather slow so if much co-ordinate drilling is planned so the modification shown at **F** and **G** will speed up matters. It is basically a sliding box to which is fixed the indicating arm. The pointer is adjusted between the plumb bob and the grid and locked there. The plumb bob will not be required again.

A depth stop A small collar with thumbscrew is fitted to the guide rod above the guide bush, **H**. If the repeating aid is used, this can be fitted for good. When each hole is plumbed, make a small distance gauge to reset the depth stop.

Wall mounted model If the space permits the wall mounted model, **J**, is the most convenient of all methods. In this case the gimbal is also arranged to move up the column as the repeating aid. Make sure to leave space for the work below the foot of the column.

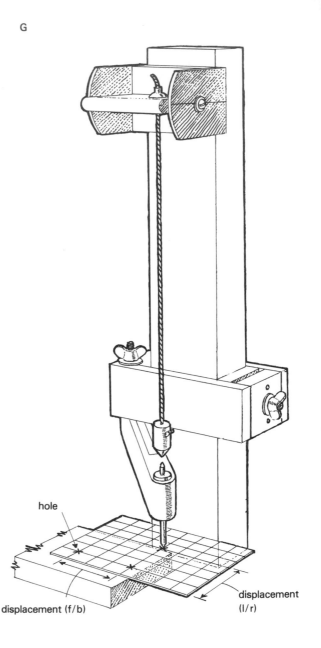

G

hole

displacement (f/b)

displacement (l/r)

Fig. 157

Edge drilling Holes in narrow-based components are best drilled on a jig shown at **K**. Make the base quite long for ease of cramping down. The workpiece is easily cramped to the block. Curved components, such as the comb or top back rail of a chair, can similarly be held by using two small blocks. Mark the centre line on the job and on the block to be sure of cramping symmetrically. Components having a curved base, such as rockers, can equally well be held, using blocks below, and measuring from the centre line.

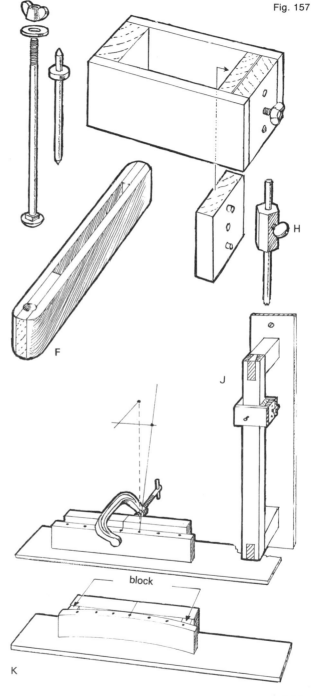

Fig. 157

block

SAWBENCH
AIDS

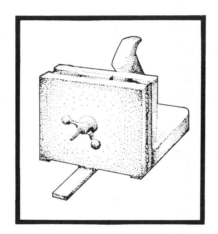

158 Safety aids for the sawbench

The circular saw and the planer have a bad reputation for safety but accidents when they do happen are invariably caused by incorrect use of the machine and a lack of safety aids. Odd pieces of wood picked up when halfway through a cut are asking for a disaster. These aids are real tools and they should be really well made. They should have hanging holes convenient to the machine and should always be returned to the same place. It is worthwhile painting these with, say, a bright orange gloss paint so that they are conspicuous and not easily lost amongst the shavings and the offcuts.

Two basic push sticks are required for the circular saw, often both in use together. A thick one, **A**, and a thin one, **B**, will cope with most sizes of work; one for pushing through, one for holding down. For rougher work a push stick ending in a short metal spike, **C**, is often recommended, but this must be used with care, keeping well clear of the saw.

Two pushers are shown for the planer. The larger, **D**, with two handles is particularly useful for thin work while the smaller push block, **E**, is useful for short pieces.

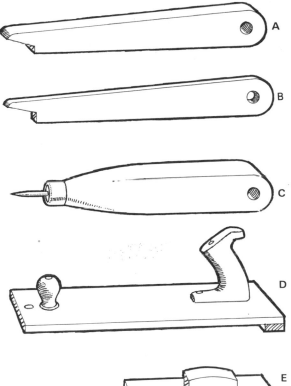

Fig. 158

Grooving and rebating on the sawbench, and rebating on the planer call for pressure in three dimensions; downward on to the table, forward producing the cut and sideways towards the fence. An assortment of push blocks for these purposes is shown, **F**, **G** and **H**. They must, however, in the case of the planer, be used with the bridge guard closed as closely as possible to them.

Safety spectacles are seriously recommended for all machine work, and that great hazard, the necktie should be removed or securely tucked away.

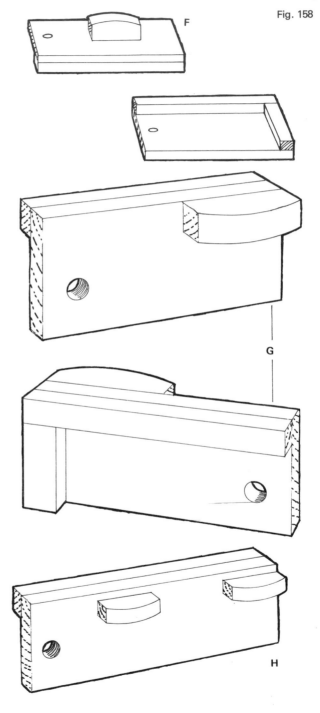

Fig. 158

159 A false table for a fine finish

The mouth of the average saw-bench is too wide to obtain a really fine finish, particularly on difficult wood. This can be obtained by making a false table of plywood or hardboard with a batten against the operator's end of the table. Having fixed the fence, the saw is wound up through the false table and the cut made. In this way splintering is avoided and a fine cut obtained, If the riving knife is in the way, the false table is fed on cutting through to the end. Such a table can be used a number of times over.

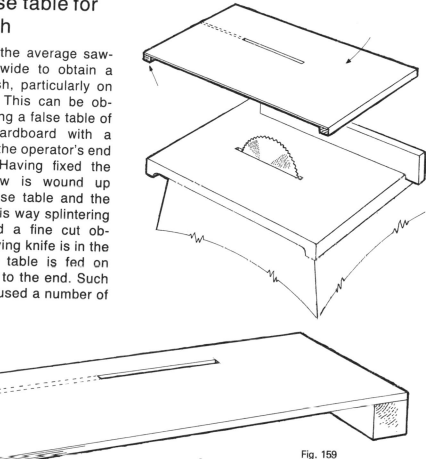

Fig. 159

160 Sawbench accessories

The fence of the average sawbench is nearly always built far too low, particularly for one common task, namely the deep cutting of say 75mm (3in) material. It is, therefore, a good practice to permanently fix a hardwood fence of about 75 × 22mm (3 × $\frac{7}{8}$in) to it extending the full width of the table. The auxiliary fences which follow fix easily to this with bolts and wing nuts, screws or G cramps.

A cramping lug Most small sawbenches have the fence secured at one end only, the operator's end. The far end can thus have a certain amount of whip. In general this passes unnoticed but when very fine and accurate work is required it can be a nuisance. To combat this a small lug **A** is lap dovetailed into the wood fence and secured by glue. When the position of the fence has been fixed a G cramp on the lug holds the far end of the fence quite firm. In deciding on the length of the lug, examine the position of any webs on the underside of the table and choose a length which enables the cramp to clear these webs in all positions of the fence.

The close-up fence It frequently happens that thin material is required particularly for the making of laminations. The small sawbench is ideal for this purpose. However, when the fence is only a few millimetres away from the saw, the guard will generally not come down. This close-up fence **B** enables the machine to be used in the normal way, with complete safety. Something similar can be fitted to quite large machines where the same problem occurs. Two pieces of well-seasoned hardwood make this L-sectioned fence. It is the full length of the table and is conveniently secured to the normal wood fence with two bolts and wing nuts. This now enables the guard to completely cover the saw and gives room also for a thin push stick to be used. A very sharp and finely set saw is necessary for the production of good laminations.

The template fence The mass production of simple flat components with straight or near straight sides can easily be accomplished with this fence, **C**. An example could be the production of a considerable number of small wooden storage boxes. The fence is similar to the close-up fence except that the horizontal member is generally wider. It can be slotted to give about 25mm (1in) of vertical adjustment or alternatively it can be secured to the wood fence by a pair of G cramps. The templates are made from hardboard or plywood and are the exact sizes of the required components. To use, fix the template fence to the normal wood fence leaving a gap of about a millimetre between the bottom of the fence and the top of the workpiece. Pin the template to a piece of the chosen material roughly cut larger than the required size. Set the saw height to just clear the fence but be sure that it will cut through the workpiece. Make sure also that the thickness of the template is such that it makes good contact with the template fence. The main fence is now adjusted so that the saw lines up exactly with the outer edge of the template fence as illustrated, otherwise the components will be either slightly larger or smaller than the template. The guard can generally be used and the work is fed in as for normal ripping. For curved work a little more care is needed. Corners are successively cut off until a complete traverse of the template can be made. A push stick is essential for safety. Concave curves cannot be attempted. Circles can be successfully cut, but if many are required the disc cutter described on page 174 is to be preferred.

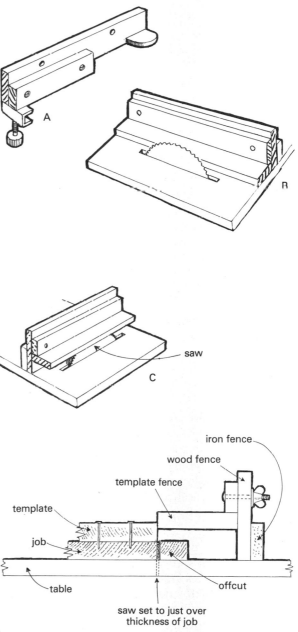

Fig. 160

147

The grooving and moulding fence Grooving and moulding using a small moulding block is now a common process on the smaller sawbenches. Without a moulding block one is restricted to merely grooving or rebating either by wobble saw or by multiple cuts with the normal ripsaw. Whichever method is used it is necessary to remove the guard and riving knife, creating a potentially dangerous situation. The set-up shown at **D** both obviates the danger and also prevents the work rising from the saw or cutter. Either of the two previous fences may be modified for use in this role or, better still, a special model may be built. In a communal workshop, with a large number of inexperienced workers, it will be found useful to prominently letter each fence with its name and so avoid confusion at busy times. One grooving fence consists of an L-sectioned piece secured to the main fence sufficiently tightly to prevent the workpiece from lifting and sufficiently slack to permit the work to slide freely. The saw is, of course, fully guarded throughout the work. The second type of fence is merely a large block, fixed in the same way, but into its underside are recessed two large ball or roller catches. This gives both adequate clearance and pressure. Work is fed through with a push stick and if it is narrow it is kept in firm contact with the main fence by the press-

ure of a wooden spring. The height of the fence is always adjusted with the cutter out of the way or fully down, to give an easy but firm sliding fit. Make sure that when the workpiece has passed through, the spring strip does not spring into the saw or moulding block.

The wide work fence When moulding, grooving, fielding or rebating on a wide board or panel it is quite difficult, with a low fence, to make sure that it stays vertical. An extra tall, broad fence, **E**, is useful for this. Multiply or blockboard is most suitable due to its greater stability. A hinged brace is best for storage. It is hinged and fitted over-long then the foot is slowly planed away until the face is truly vertical. A suitable notch is cut to straddle the existing fence. Screw holes can be arranged to match those already in the hardwood facing on the original fence.

Alternative fence face for moulding block Frequently when the moulding block is used either for moulding or rebating, the full width of the cutter is not required. Part of it is masked behind the wood facing of the fence. The higher the cutter is raised, the further the unrequired portion cuts into the wood. By the time several moulding operations have been completed, the wooden fence has been considerably cut away. If thin pieces are later fed through for normal ripping, they are liable to catch where the wood has been cut away. It is essential, therefore, to keep a fence, **F**, specially for this type of work so that the main fence remains intact for its original purpose.

Straight-edging A long board with an untrue edge can be trued up by one of two methods, **G**. First pin or fix with glue dabs and paper joints, any convenient width of straight-edge, to the line required. Fix on the template fence, adjust the saw height to just cut through then adjust the template fence to just clear the saw. Operate as described under the template fence. The second method is to similarly fix to the required line a straight-edged piece wider than the offcut. This time it is fixed to the right-hand side of the line. It must overhang the board everywhere and its true right-hand edge must be parallel to the line required. The correct saw height will just cut through the workpiece and slightly mark the straight-edge. In use the straight-edge runs against the normal fence. These methods are particularly useful when a large number of very rough pieces are being prepared for sawing out to size.

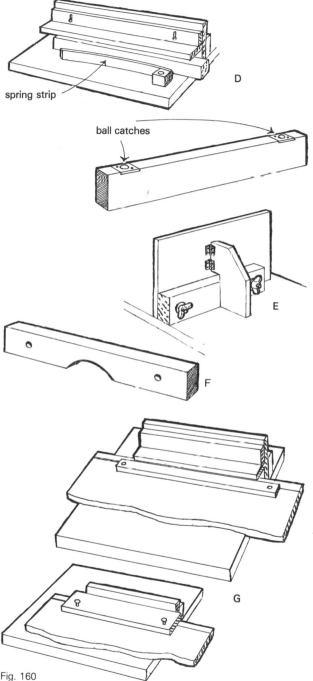

spring strip

ball catches

D

E

F

G

Fig. 160

Pattern sanding The requirement here is similar to template sawing, in fact by substituting a sanding disc for the circular saw, the template fence can be used. Another method, **H**, is to make a false table from blockboard and lip its edge with a thin strip of metal standing up about 3mm (⅛in). A steel bar is fixed beneath to go into the sawbench groove. The lip is positioned as near as can be arranged to the sanding disc. This time the pattern, from plywood or hardboard is held downwards and continuously up to the metal fence. The patterns must be cut appropriately undersize.

45° sawing J. Pieces of this shape are sometimes required, notably for glueing blocks. Not all benches tilt, and even on those that do this method is more comfortable to use. The fitting is rather like a rounding cradle. It is screwed to the main fence which is then positioned as accurately as possible with the saw wound down. The saw is then wound up through the fitting to coincide with the centre of the V cut. A little sideways clearance on each side is preferred. The work is fed through in a straightforward way using a push stick. Except in the smaller sizes the guard will probably have to be removed. Alternatively, glue together with a thin plywood insert, leaving a space for the saw. In addition a sawdust groove can be arranged. The

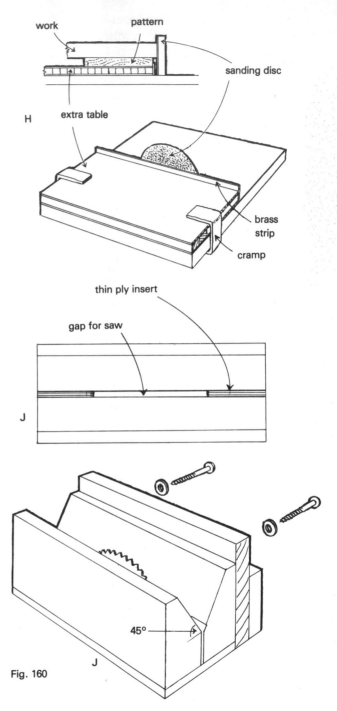

work   pattern

sanding disc

H   extra table

brass strip

cramp

thin ply insert

gap for saw

J

45°

J

Fig. 160

150

workpieces will then fit more perfectly into the jig.

Multiple cross-cutting A common error and a dangerous habit is to use the main ripping fence for cross-cutting. An offcut nestling between the saw and the fence needs only a small amount of rotation to reach a position where the length of its diagonal is greater than the distance between saw and fence. In this state the offcut binds and is flung out violently, possibly breaking or buckling the saw. To prevent this from happening, a block is bolted through a hole in the main fence and the fence is adjusted to compensate for this, **K**. If the thickness of the block is made as a round number, for example 50mm (2in), then this must be added to the setting read on the sawbench scale. In use the workpiece is fed up to the block, held very firmly to the cross-cut slide and passed over the saw. The offcut now has plenty of space to fall out into and the possibility of binding, with its dangers, has been eliminated. Used in this way as a length stop, many identical pieces can be cut.

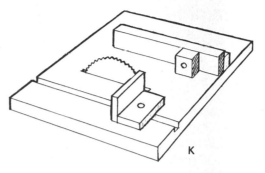

K

161 A straight-edger for the circular saw

This enables a piece of wood with a waney or untrue edge to be sawn straight and then, subsequently, cut parallel again. A baseboard of blockboard or multi-ply is fixed to a metal bar sliding in the sawbench groove in such a way that when passed over the saw a sliver will be taken off the base. This gives a true and straight edge from which the set-up is arranged. Draw a line on the workpiece where the cut is required. Place this against the true edge of the aid and tighten up the two cramping devices. Now simply pass over a saw, set approximately to the required height, and a good straight edge will result.

162 Tapers on the sawbench

A basic taper jig is shown at **A**. Plane a length of hardwood to the amount of taper required and fit a handle and a push block. If this latter is angled slightly as shown it will help to stop the workpiece from kicking up. A plan view of the sawbench with the jig in use is shown at **D**.

A strip from this jig can be used to make the vice jaw in Fig. 4 which is used to hold the tapered component for subsequent planing.

152

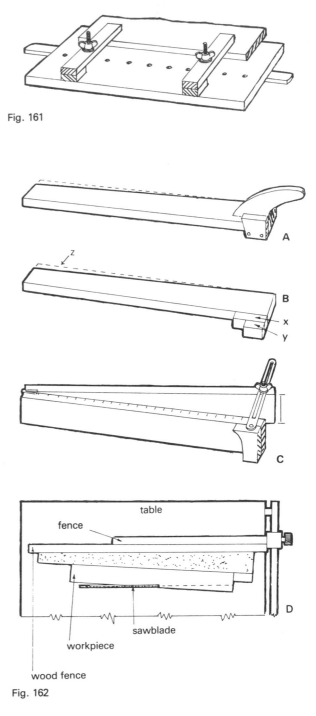

Fig. 161

Fig. 162

A variation for producing, for example, legs tapered on all four sides rather than more commonly the two inner faces is shown at **B**. The first two tapers are given by **z**, the leg being pushed by the block **x**. For the second two tapers **x** is used as the spacer and **y** is the pusher. A full-sized drawing is recommended for this model.

An adjustable taper jig is shown at **C**. Two hardwood strips are hinged at one end. Their height must be just greater than the height of the sawbench fence. The push block and slotted metal clamp bar are quite clear. One strip is calibrated from the hinge centre in millimetres or inches. The amount of taper, for example 20mms in 400mms can thus be easily fixed.

163 The tenon tool

There can be as many as eight cuts to make a tenon so this tool is, in fact, a tenon cheek cutter. Tenon cheeks have always been cut on circular sawbenches but a glance into one of the smaller commercial workshops often will reveal practices guaranteed to make the blood run cold! Here is a very precise and safe way by which tenons can be cut, even by the inexperienced. I devised this method of work to complement mortises already being cut by machine.

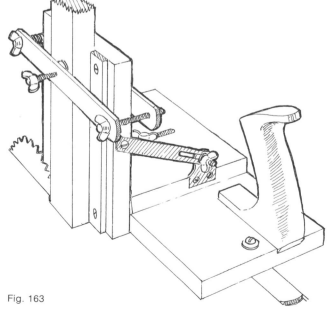

Fig. 163

To cut tenons by cramping to a sliding vertical face is common knowledge and angled tenons required for chair-making, etc., meant that adjustment to angles other than 90° was obviously necessary.

There can be much debate as to which side of the saw one works, I have chosen to work on the right-hand side of the saw, from the working position, using the left hand and a close guard on the left.

Precise sizes for this tool cannot be given because of variations in sawbenches and individual requirements, but where helpful, recommendations are made. Any hardwood, multi-ply, or blockboard is suitable. A baseplate about 21mm ($\frac{7}{8}$in) thick is fitted with a plane handle and a metal slide bar, **A**, to suit the sawbench groove. This slide is fitted with two screws, one being a countersunk woodscrew from beneath, the other is a metal screw from above, **C**, for which the slide bar is tapped. The metal screw works in a slot or enlarged hole for accurate adjustment later. The long edge of the base ought to be about 50mm (2in) from the saw depending upon the average thickness of the work to be undertaken, and an extra, alternative pair of holes might be preferred.

A sliding plate is fitted to the base. This and the base can be grooved on the circular saw at the same setting. The separate slides of hardwood or metal, **D**, are fixed permanently into the grooves in the base. There is a centre slot, **E**, and securing bolt **B** with knob or wing nut. The edge of the sliding plate should be slightly angled. The working face is hinged to the sliding portion and the tool now assumes a purposeful appearance. Two small plates, mirror plates are used here, **F**, are fixed to the sliding member and two slotted bars, **G**, connect to the working face. It is now possible to secure the working face at exactly 90° or any other angle. Set the 90° with a set square rather than with a try square as the latter tend to become worn by the marking knife and untrue by being dropped.

A stop piece, **H**, is now fixed to the working face to prevent the work from kicking back when sawing and several alternative sets of holes, **I**, should be provided to cope with the varying widths of workpieces.

The clamp bar, **J**, can be made of wood or metal, bearing in mind that no great pressure is required. A convenient refinement is a pair of coil springs, **K**, which will allow the clamp to open as the wing nuts are slackened. The work is held by a convenient clamp screw, **L**, but if metal

Fig. 163

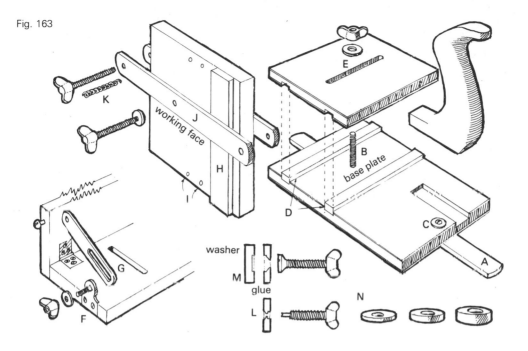

working facilities are not available, a wooden clamping foot can be made up as illustrated, **M**.

Adjust the face so that it becomes parallel with the saw. To test the accuracy of this, gauge the end of a wide piece of wood and re-adjust as necessary until the tool cuts exactly to the gauge mark. Without further adaptation the tool will work making two cuts for each tenon, but it will be better if two saws can be used to make both cuts simultaneously. Most sawbenches have a spindle long enough to cope with a tenon up to 10mm ($\frac{3}{8}$in) but, of course, some readers will have to be content with taking two cuts. Buy two identical ripsaw blades and

keep them sharpened to the same diameter.

To use two saws a system of spacing washers is needed, **N**. I have four thicknesses made to $\frac{1}{8}$, $\frac{9}{64}$, $\frac{5}{32}$ and $\frac{11}{64}$ in and several washers in each thickness which, in combination, meet every requirement. Wooden spacers can also be used and various thicknesses of plywood, alone or in combination, will prove successful. Plywood varies quite a bit within nominal sizes but it can be veneered either on one side or on two sides so that a fine degree of adjustment can be obtained. It is worth recording the various spacers used in combination with particular mortise chisels. The spacers required will vary according to the amount of set on the saws at any particular time and the amount of use they have had.

A safe and convenient guard is essential and can be made from hardwood and acrylic plastic, **O**. It is screwed into holes tapped in the saw table and the slots allow it to be positioned as near as convenient to the workpiece. The securing screws, when not in use, are parked in tapped holes in the hardwood base, **P**.

The method of setting up is as follows. First, check the vertical face. Put in two saws with spacing washers estimated to suit the mortises, which are always cut first. Remember that because of the set, a 6mm ($\frac{1}{4}$in) spacer will not produce a 6mm ($\frac{1}{4}$in) tenon; it would be very convenient if it did. Fix the guard to suit the thickness of the workpiece, then make a practice cut in a piece of waste wood. Cut off the shoulders and try it in the mortise. Continue to adjust the spacers until a good fit is obtained. Remember that if too tight a fit is accepted most of the glue will have been rubbed off the tenon by the time that it has been cramped up into position. Record this spacing for future work. Adjust the sliding member until the saws are accurately cutting to the gauge lines. Lock when correct and then adjust the height of the saw. Finally, fit the stop bar

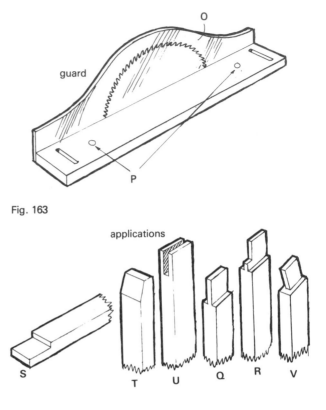

guard

Fig. 163

applications

156

so that the clamp foot is central on the workpiece. In working stand well to the right, pushing the tool with the left hand, as in this way the sawdust throwout is largely avoided. Go through the saw and then stop and switch off. This makes a better cut than if the work is brought back across the saw and makes sure that the workpiece is never changed when the machine is running. The true face of the work must always be positioned against the workface of the tool. Wear an eyeshield and remove loose clothing such as a necktie for safety. Other applications of the tool are the cutting of half-lap joints, bridle joints and angle cuts as shown at **S–V.**

164 Shoulder cutting tool

The sawing of shoulders on the circular saw bench is not such an easy task as would appear on first consideration. Most owners of sawbenches have, at some time, tried it. Unless a really satisfactory jig is used, the machine-sawn shoulders are invariably inferior to shoulders sawn by hand. Using merely the cross-cut slide, the work tends to drift away from the saw and the matching-up of opposite shoulders with real accuracy is well-nigh impossible. If the workpiece is held against the rip fence, a dangerous situation is created in which the offcut cheek can bind

between the saw and the fence and then hurtle out with considerable velocity, maybe damaging the saw if not the operator. The average cross-cut slide with protractor is small and not very precise so if its right-angle is not perfectly set, a twisted pair of shoulders will result.

This tool does a successful job and can repeat it as often as required. There are two pre-conditions; one is that the ends of components shall have already been cut squarely and accurately to length and the other is that the sides of the components be accurately parallel. As we are considering machine work, these requirements are not unreasonable.

The holding of the components on a travelling base helps to reduce the tendency to slide away from the saw, consequently this baseplate is the first component to be made. Any stable hardwood, blockboard or multiply is suitable and its length should be approximately 300mm (12in) and width a little over twice the distance between saw and saw groove. As in previous tools in this series a steel slide bar is required, a good sliding fit in the saw table groove and a little longer than the baseplate. One end is secured from below with a countersunk wood screw. The other end is bored, tapped and secured from above with a round-headed machine screw with a washer. This operates in an over-size hole, thus allowing adjustment of the jig in relation to the saw. A handle for a wooden jack plane is housed into the baseplate, glued and screwed from below, positioned so as not to interfere with the screw securing the slide bar. The slide bar is positioned as accurately as possible, parallel to the side of the baseplate. The baseplate is now passed across a sharp circular saw, thus providing its working edge. The fence is screwed and glued at an exact right-angle to this working edge.

Before fixing, cut out a wide groove in the top of the fence. This could be cut out by many cuts on the sawbench and then smoothed up with a router. Two countersunk machine screws come through the fence, and if they are put through tapped holes they will not turn when in use. A slotted bar slides in the groove of the fence, and it may be of 10mm ($\frac{3}{8}$in) plywood, or of solid stuff. A 6mm ($\frac{1}{4}$in) slot is cut in this or, if of solid wood, it may be built up with a resin glue. Plan the length of the slot to suit the maximum length of tenon likely to be encountered, wing nuts with washers holding the slotted bar in place. A length stop is glued to the end of the bar and if it is glued *in situ* on the fence and base, the end stop will be exactly parallel to the sawn side of the

base, that is parallel to the saw.

Test the accuracy of the tool in the following way. Wind up the saw to a little above the base-plate height and adjust the length stop to about 13mm ($\frac{1}{2}$in). Place a wide piece with a true edge against the fence. Hold it firmly against the fence and length stop and pass over the saw. With an accurate try square test this cut. A more precise way of testing the right-angle is to cut half-way through a wide, parallel piece, turn it over and repeat, **A**. Any error will be obvious. Working in softwoods requires less accuracy than hardwoods as the former will crush up a little. A slight adjustment may be necessary at the slide bar.

The tool is now ready to cut shoulders of tenons on which all the other cuts have already been made. The shoulder is always cut last and a sharp cross-cut saw must be used. Adjust both the saw height and the end stop position accurately on a prac-tice piece. On some saw-benches it is possible to operate with the normal guard in pos-ition and, as the cheeks fall off, poke them clear with a stick, *not with the fingers*. It is danger-ous to leave a heap of them round the saw.

By modifying the saw height but not the end stop, it is possible to cut the set-in. Saw height and

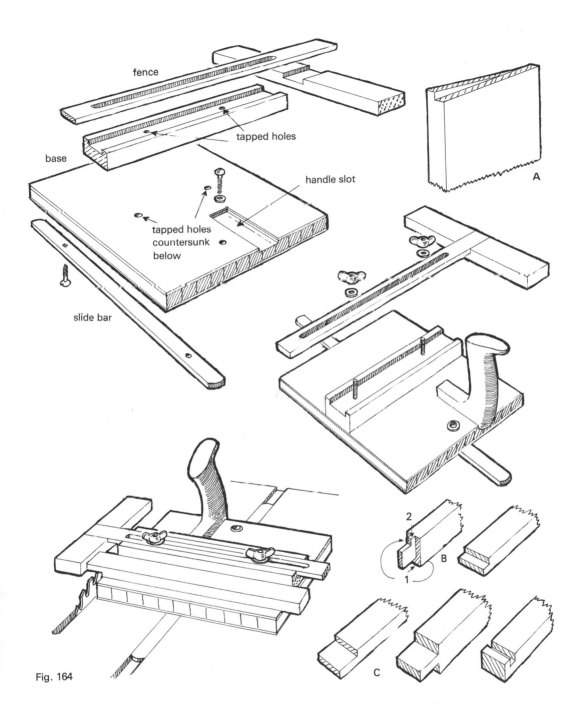

fence

tapped holes

base

handle slot

tapped holes
countersunk
below

slide bar

A

B

C

2

1

Fig. 164

159

end stop will need to be altered to enable a square haunch to be cut. Frequently a tenon is not set centrally on a rail. In this case, put through all the pieces with the true face down then, having altered the saw height, repeat with the true face up. The tool will also help in cutting half lap and bridle joints. No doubt the reader will find still further applications. When it is required to cut shoulders at angles other than 90°, prepare an accurately-tapered block and insert this between the fence and the workpiece.

Finally, always check the right-angle before starting an important run, as a knock may have upset the setting of the slide bar.

165 The haunch tool

The tenon tool Fig. 163 is, more precisely, a tenon cheek cutting tool. In mechanised tenon cutting what is now wanted is a similar tool, cutting at right-angles to the cheeks. This will cope with the haunch, the set-in, stub tenons (pinnings) and anything of a like nature. The solution in this case is fairly simple. No sizes, but a few recommendations are given as the user's requirements will vary according to the sawbench available and the type of work likely to be undertaken. The material from which the tool is made is not of great importance; beech is ideal but blockboard or multiply can be used.

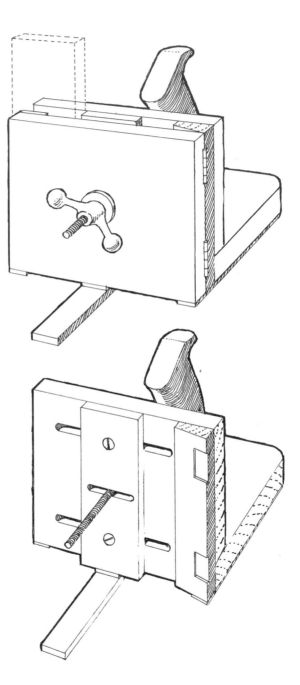

Fig. 165

160

A working face is dovetailed or otherwise jointed to a base, which in turn is fitted with a steel runner to fit into the saw table groove. This runner is fixed by two screws, a countersunk woodscrew from beneath and from the top by a metal screw into a threaded hole in the runner. This latter should be through an oversize hole, giving a sloppy fit, thus allowing later for a fine adjustment ensuring that the working face is an exact right-angle to the saw. The end of the working face should just clear the saw when sliding past. Friction is reduced by running on two narrow strips of laminate. A further strip must be put under the slide bar if this is not to protrude above the table top. Any rock is eliminated by grinding on abrasive paper taped to the sawbench table.

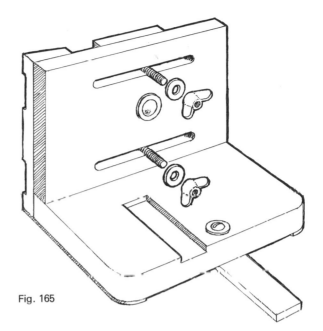

Fig. 165

A positioning stop for the work is required to make sure that components produced are identical. The working face is slotted to take two brass screws. The stop is drilled and tapped for the two screws which pass through the slots in the working face and secured on the far side with wing nuts and washers. This arrangement makes sure that the screws do not revolve when the wing nuts are tightened. The handle for a wooden plane is housed and glued into the base. Take care not to choose a position which is inconvenient for handling the wing nuts or the screw securing the steel runner. In this form, the tool is usable, the workpiece being held by a G cramp.

There is always a shortage of cramps and also there is the danger that a cramp may vibrate loose. This prompted the design of a holding device. A block thicker than the average work is glued to the end of the working face furthest from the saw. On to this is hinged a flap clamp, and to make sure of a good grip, a strip of thick rubber is glued on to the working end. A tight grip is achieved by one of those quick threaded wing nuts with screw. The hole for the screw in the hinged flap needs to be extended into a slot. A coil spring made from piano wire, used in model aircraft, eases operation by making the clamp open when the nut is slackened.

Operation is quite straightforward. The height of the saw is adjusted to just clear the marked line. The positioning stop is secured and the workpiece clamped in. For a right-handed person it is safest to operate in the left-hand groove of the saw-bench, seen from the operational position. The guard designed for the comb jointer, Fig. 169 is equally successful with this tool.

To hold very wide pieces it will be necessary to remove the clamp, clamping screw and the positioning stop.

166 Mitre-cutting tool for tenons

First of all it must be decided on which side of the saw to work. I have chosen to work on the left as I did for the shoulder cutting tool, but this tool can equally well be built to operate on the right.

As previously, the baseplate **A** is made first, and almost any material is suitable. It should be about 50mm (2in) wider than the distance from the saw to the far side of the saw-bench groove. It must be made truly parallel first, then one long edge should be sawn to exactly 45°. The steel slide bar **G** is fixed at the front edge by a countersunk woodscrew from below, and at the handle end by a round-headed machine screw. Tap the bar for this and arrange to pass the screw through a large hole in the base using a big washer. Fine adjustment is made here and the bar is fitted later.

The working face **B** is built up from two layers of plywood, glued together, and in this case 8mm ($\frac{5}{16}$in) plywood was used. The top layer is made of three pieces. Only two, the fixed pieces, are glued but the unglued piece, which is the sliding portion **D**, is held in place during the gluing thus giving an accurate sliding fit later. Naturally, when the cramps are on, this piece is removed. A slot is cut in the centre of the groove which takes the screw in the sliding portion. A ¼in Whitworth countersunk machine screw passes through a threaded hole in this part, then passes through the slot and is secured from beneath with a wing nut and washer.

Having removed the sliding part, glue and pin the working face to the base. Check the angle with a 45° set square. After gluing it was found that the joint was quite rigid enough for normal use, but bearing in mind possible misuse and damage by dropping, it was thought wise to re-inforce with two 45° plywood brackets, **F**. A wooden plane handle or a turned knob is then fitted, taking care that it is not going to interfere with the adjusting screw into the slide bar. Now position the steel bar. The vertical stop piece or fence **C** is screwed into place, but not glued until it has been tested and found to be correct. Next fit the sliding member in place and tighten the wing nut. A small hardwood or plywood strip, about 3mm (⅛in) thick is glued to the bottom edge, making the horizontal stop piece **E**. Test with a set square to ensure that it is exactly at right-angles to the vertical fence.

To test the tool, gauge a mark on the end of a wide piece of wood. Put this on the workface, having removed the sliding stop, cramp and make a cut along this line; if the cut is not truly parallel adjust the steel bar until correct.

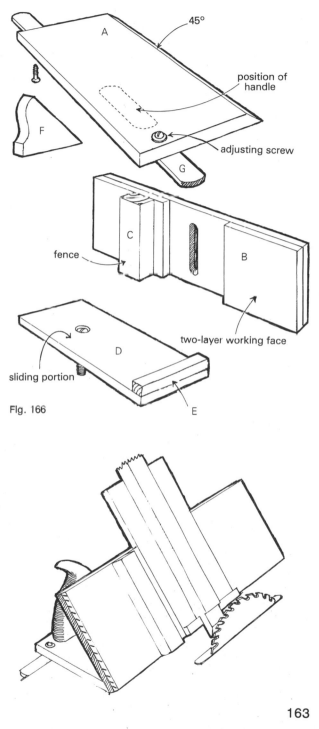

Fig. 166

163

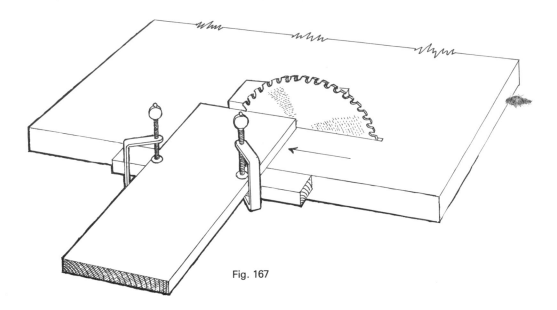

Fig. 167

Now for actual use. Put a sharp cross-cut or combination saw in the machine. Put on the workpiece and from below adjust the sliding stop until the mitre is being cut in the correct place. Adjust the saw height until the saw just protrudes. The easiest guard to make is just an angle strip, which is screwed to the rip fence of the sawbench and is adjusted to be as close as is convenient to the work. There is no need to cramp the workpiece as the forces at work will force it against the two stops. If the sliding portion is removed and turned over, that is, with the stop out of action, the tool can be used for straightforward 45° cuts, the workpiece being held by a cramp. In this form a number of other joints and uses will be found possible.

167 Squaring ends on a small sawbench

A strong batten, cramped square to the board, will slide easily along the edge of the table permitting the ends of comparatively long boards to be squared. For pieces of any length this is a much safer and more efficient method than using the small cross-cut slide.

168 The cross-cut table

There is no doubt that small sawbenches are sometimes expected to do work for which they were never intended. The cross-cutting of a wide board is a case in point. This task, if done by machine, requires either a pull-over radial arm saw or one of the older pendulum saws. Providing, however, that the lengths are not excessive, this device makes a very fair job, safely. The puny cross-cut slide is quite useless in this case as a wide board cannot be held in firm contact with the small fence. There is also a natural tendency for the wood to slide or drift sideways, away from the saw. The remedy is a complete sliding table. There are two choices; one is a table completely covering the saw table, with its own safety guard – a bulky item calling for a messy setting up. The other described here is a half-width table, which obviously is not only smaller but can slip on or off the saw table, using the existing guard system with no inconvenience.

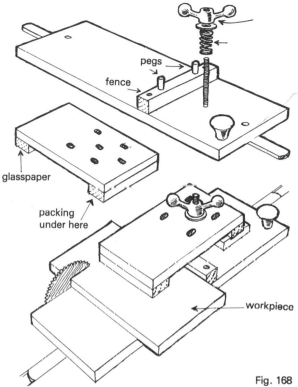

Fig. 168

The baseboard can be made from blockboard and it should be considerably longer than the actual saw table and, in width, just over the distance from the saw to the left-hand edge of the table. I have chosen to operate on the left of the saw but the design is quite reversible.

A piece of bright drawn mild steel to suit the sawbench grooves is screwed underneath with two screws, three will distort if not perfectly in line and that means binding in the groove. Fit this steel slide in such a position that a pass across the saw will cut a very thin strip from the base, thus giving a working edge parallel to the saw. A turned knob or a plane handle is fitted to push the tool and a hole bored at the far end serves to hang the tool when not in use to prevent knocks which maladjust it.

The fence is screwed to the base in a position suitable to the particular sawbench. Each screw has a washer and one of the holes in the fence is elongated to a slot for adjustment later. For a start, fit this square with the working edge of the base. Insert two dowel pegs into the fence as shown. The clamping device is simplicity itself and operates like

165

a number of my jigs, on one of those old-fashioned fast, square-threaded screws with big wing nuts, alternatively use a wing nut with a conventional thread. The screw comes up from the base and over it fits the clamping jaw. This is of 21mm ($\frac{7}{8}$in) hardwood close in width to that of the base. It is thickened up with a strip at each end to bridge the fence. Two holes are made for the clamping screw, giving two positions of the jaw, for wider or narrower wood. A few strokes of a round rasp elongate these holes slightly. Similar holes near the sides fit over the short dowels in the fence which prevent the jaw from twisting sideways. The business end of the jaw is slightly rounded and on to it is glued a piece of coarse glasspaper. Under the other end goes a wood block, approximating in thickness to the workpiece. A piano wire coil spring over the screw opens the jaw when the nut is slackened.

The operation is fairly obvious. Take away the rip fence. Put in a sharp cross-cut or combination saw and find a packing piece of suitable thickness. Check that the standard guard is going to operate satisfactorily and choose that hole in the jaw which will put the clamping pressure nearest the centre of the workpiece. Adjust the height of the saw to be just over the combined height of the job and the tool. Clamp in the

workpiece and cut off the end with some to spare. Check the squareness of this cut and, if necessary, adjust the fence. Taking a preliminary nibble, adjust the workpiece until cutting to the line, then cut through. In the cross-cutting of wide and sometimes longish pieces, each piece must be checked separately before sawing. In this respect the tool is not automatic.

169 The comb jointer

The most perfect method of jointing together boards, as distinct from legs and rails, is one of the variations of the dovetail joint. Alternatives range from the crude nailed joint to the cross-grained tongued joint and all of them are inferior. Industry, by the use of expensive machinery using multiple cutters, came up with the comb or box joint. Using modern adhesives, this is almost as strong as the common dovetail and what it loses in design efficiency it makes up for in accuracy of fit. The joint has considerable decorative possibilities. Here is a tool which will cut this joint on any small sawbench, in this case a 200mm (8in) saw. The construction and operation are quite straightforward and success is guaranteed.

The tool is made from any stable hardwood, in this case beech. The face and the base are jointed at an *exact* right-angle. A lap dovetail was used but there are alternatives. The exact width of the tool is governed by the saw-bench to be used. The saw lines up with the centre of the face and the ends must extend at least 25mm (1in) beyond the saw-bench grooves. If there is only one groove, make the tool symmetrical about the saw. Two 10mm ($\frac{3}{8}$in) holes are bored centrally over each of the grooves. Two strips of bright drawn mild steel are cut which exactly fit and slide well in the saw table grooves. It is important that these strips project in front of the face to prevent the tool from tipping forward in use. Both strips are drilled and tapped with 6mm ($\frac{1}{4}$in) threads and are secured to the base with round-headed screws and big washers. The very sloppy fit of these screws in the base allows accurate adjustment of the face, which must be square to one runner and hence square to the saw. When placed *in situ* with slack screws the second guide finds its own position. Once it has been established that the tool will slide sweetly over the table, the slides can be removed until later.

A 25mm (1in) groove is cut vertically on the face to a depth of about 8mm ($\frac{5}{16}$in). This takes the shear block and can be most conveniently cut on the saw-bench. Similarly, a rebate is cut at the bottom edge to take the mechanism. Additionally a small groove is worked to take a nut. It was found easier to cut the rebate to the full length and then fill in the part not required. The tool can be faced with plastic laminate and if this is done it must be included when considering the depth of the rebate. Put an over thick piece in the shear block groove, glue the plastic laminate and push it hard against the block so that there will be no need to trim the laminate at the groove edges.

As the saw will cut a groove through the base, a strengthening piece has been glued over this area. Into it is housed a bought jack plane handle. The handle should not be set centrally but just short of the centre line, on the left when viewed from the operating position. This allows ease of access for the screwdriver to the shear block screw, the hole for which, 5mm ($\frac{3}{16}$in), may now be drilled.

To reduce friction, two small strips of constructional veneer or plastic laminate are glued to the base. They must also be under the steel runners otherwise these would not fit flush with the table top. Check now that the tool stands without any wobble. If it rocks on the table, rub off the high spots on a sheet of abrasive paper taped on to the saw table. Finally make a hardwood shear block. It must be a good fit in the groove, flush with the face, but it may be a little taller than the face

Fig.169

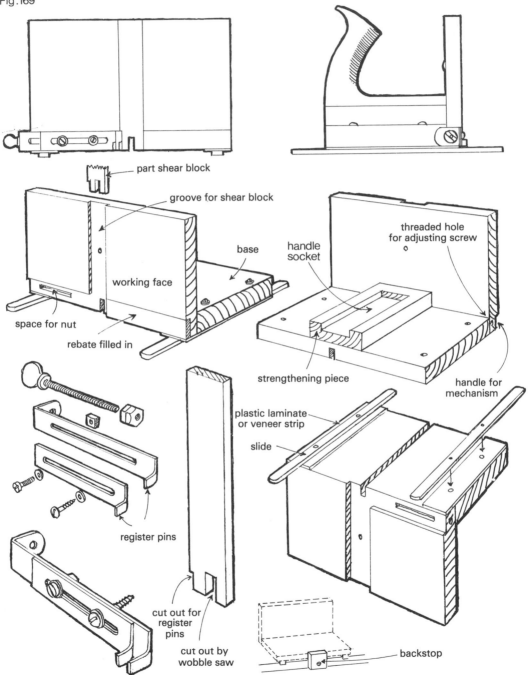

part shear block

groove for shear block

base

handle socket

threaded hole for adjusting screw

working face

space for nut

rebate filled in

strengthening piece

handle for mechanism

plastic laminate or veneer strip

slide

register pins

cut out for register pins

cut out by wobble saw

backstop

as it gets progressively cut off in use.

We now turn to the mechanism and here also the sizes and details may be varied according to the materials available, and also to suit the distance between the saw and the groove. This one was made from brass strip which measured 20×2.5mm ($\frac{3}{4} \times \frac{3}{32}$in) but steel or alloy, if not too soft, would be equally suitable. The main plate is bent to make its register pin.

With the pin touching the saw, the second bend must be made about 12mm ($\frac{1}{2}$in) beyond the wood base. The smaller plate has only one bend to make the register pin, and any excess material must be cut away when making them. Also, their height above the table must be less than the thinnest material for which it is contemplated using the tool, yet, at the same time, maximum strength must be preserved; so think twice and cut once! Slots are cut as indicated and then the two plates are attached to each other by a cheese-headed 2BA or similar metric screw and nut, the nut being soldered into position. The plates will be held on to the tool with a round-headed 1in by no. 8 woodscrew. The distinction between screw heads makes for convenience in adjusting.

The adjustment of the plates is by means of a shouldered thumb screw $2 \times \frac{1}{4}$in BSW but anything available with a similar thread would be suitable; if metalworking facilities are available, a knurled knob could be made. Two thin lock nuts secure the screw to the main plate. The wooden base must be threaded to take the adjusting screw. If you are fortunate to have a threaded brass insert it can be used or a brass nut may be let into the wood, but in practice, if a good hardwood has been chosen and a hole in it threaded to full depth, it will take a great amount of use before the wood thread begins to strip. With the refitting of the runners the main tool is now complete.

A pair of wobble washers is now required to produce the wobble or drunken saw. If the sawbench is provided with a proprietary adjustable wobble saw, this is perfectly suitable. It should be pointed out, however, that these saws seldom run in the same position as the normal saw, so this must be borne in mind in the initial planning.

As an alternative to the complete wobble saw unit, sets of adjustable wobble washers are commercially available. These are calibrated and give a great range of widths of cut and are available with bores of $\frac{1}{2}$in, $\frac{5}{8}$in and $\frac{3}{4}$in. They are economically priced and quite satisfactory apart from having very sharp edges, which a

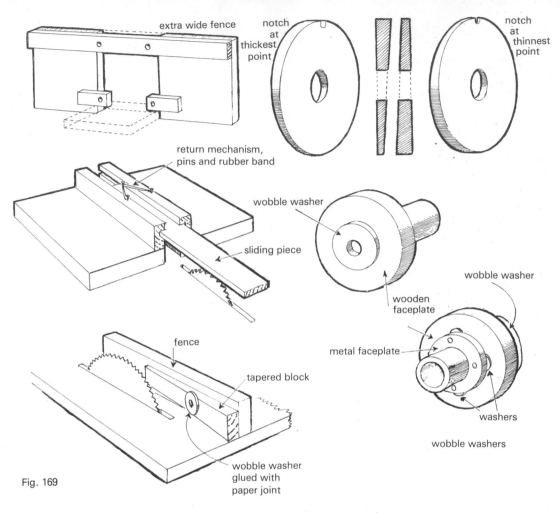

extra wide fence

notch at thickest point

notch at thinnest point

return mechanism, pins and rubber band

wobble washer

sliding piece

wooden faceplate

wobble washer

metal faceplate

washers

wobble washers

fence

tapered block

wobble washer glued with paper joint

Fig. 169

touch of emery cloth soon corrects.

For those who cannot obtain either of these alternatives, wobble washers can be made although they can be a problem, the solution of which depends upon available equipment and the co-operation of friends. It is advisable when making them to make several pairs of washers giving differing widths of pins in the comb joint. A slope in the sectional view of about 1 in 50 is a fair average so two further pairs one with more slope and one with less slope would satisfy most needs. An engineer will find little difficulty in making these in metal but if wood has to be used here are two methods.

If a lathe is available, glue a

170

hardwood square with a paper joint on to a wooden faceplate. Turn a disc and bore out the centre hole. Slacken the screws holding the wooden faceplate to the metal faceplate and insert washers between the metal and wooden faceplates of two screws only. Screw up all the screws tightly. The disc will now run unevenly. Turn the face flat again, then break off. Repeat to make the pair. Further pairs are made by altering the thickness of the washers.

To make wobble washers without a lathe, prepare hardwood blocks over-thick, cut circular and bore out the hole. Next prepare a stout block which has been tapered to the chosen slope and glue the disc on to it with a paper joint. Pass the block with its disc past the circular saw, thus cutting the disc to the required tapered section. Break off the joint and repeat for the other washers. File a notch at the thinnest part of one washer and at the thickest part of its pair. These must line up when the saw is sandwiched between them. The diameter of the saw and the amount of set on it, in addition to the slope of the washers, determine the width of the saw kerf. It is necessary to experiment, therefore, rather than copy given dimensions, so while the jig is set up, make a sufficient number of pairs of washers, identifying each pair clearly.

The chosen saw, a ripsaw, will require an indicating line to line up with the notches on the washers. The mouth of the sawbench may not give sufficient clearance for the wobble saw. If not make a suitable mouth from alloy or brass.

While undoubtedly a number of workers will be quite happy to run the saw without a guard, this is strongly to be discouraged, a guard is essential. The one recommended is both easy to make and effective and it will serve for other tools as well.

The base is screwed to the saw table which has been drilled and tapped for the purpose. Two grooved blocks are glued to the base and a plywood strip guard slides in these. The work will push the guard forward and two rubber bands or light springs will return it, so that the blade is always covered. The guard stops forward movement but some means must be devised, suitable to the sawbench, to stop backward movement. When not in use, this stop may be stored screwed to the base of the tool.

Operating instructions The comb jointer, because of its mathematical nature imposes certain limitations on the design. Assuming a box-like structure is being planned, whereas dovetails would be arranged to fit the size of the box, when comb jointing, the box must be made to fit the joint unless part of a pin somewhere is acceptable. Plenty of offcuts of the chosen thickness are required for the setting up.

Fit a sharp ripsaw between chosen wobble washers, line up the notches and put in a suitable mouth. Spin the saw by hand to make sure that it is not fouling anywhere. With one of the offcuts set the saw height. There are two choices; set the saw low and clean the job down to the joint. This way cramping is easy. Otherwise set the saw high and clean the joint down to the job. This is nicer but necessitates the making of comb-shaped cramping blocks, so on balance the first method is preferable. The amount by which the saw is set low is quite minute.

Hold an offcut against the fence and up to the register pins. Make a cut, producing one notch. Unlock both locking screws then, by inserting a screwdriver between them and gently twisting, make the two register pins a tight fit in the cut, **V**. Tighten the cheese-headed screw and do not alter it again. Tighten the round-headed screw and take a few more cuts placing each new cut on the register pins to make the next one. Repeat the row of cuts on another piece and try the joint, **W**. To adjust the fit, slacken only the round-headed screw and turn

Fig. 169

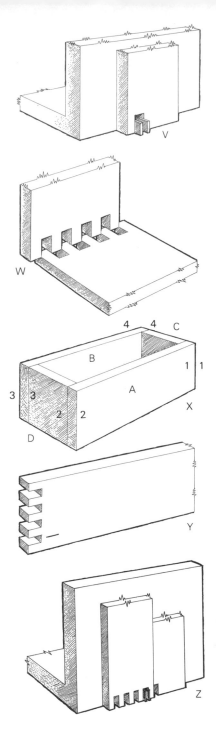

172

the adjusting screw, then relock. Continue this adjusting until the joint fits at the complete width required. Hardwoods need to be a looser fit than softwoods. If the settings are not disturbed and the saw is lined up with the notches on the washers there is every chance that on a subsequent operation, no further adjustment will be required. When a perfect joint has been achieved the shear block is fitted or sawn off and refitted to eliminate the 'rag' on the finishing side of the cut. Notch the shear block round the register pins then fix it in place with the screw previously mentioned. Pass it once over the saw. Now fix the guard and the back stop.

Push the box together as illustrated with the true faces inwards and the true edges down. This means that the poorer looking sides are the true faces. Letter the sides A, B, C and D, A and B being the longer sides. Number the joints 1.1, 2.2, 3.3 and 4.4, as at **X**.

Work commences on the long pieces. Put A on the fence, true edge *against* the register pins and make a cut. Mark this cut with a pencil then place it over the register pins and repeat. Continue in this way until the whole width has been covered, **Y**. Note that the true edge is still leading. Exactly the same is done to the ends of B. The long

sides are now jointed. The short sides have the opposite jointing and this is arranged as follows.

Put the first cut made on A, marked with a pencil, on the register pins *with the true edge facing left* from the operators viewpoint, i.e. facing the saw. Now place piece C on the face, true edge leading and butting firmly against A. In this position pass over the saw, **Z**. Piece A can now be put away. C will be seen to have a one-sided cut. Push this against the register pins then work as before. Repeat on the other end of C. Similarly cut both ends of D. Remember that the true edge leads and that short sides start by butting against A, whose first notch is over the register pins.

The jointing is now complete. Polish the insides before assembling protecting the joint from wax or other polish with masking tape. Though a good joint should, in theory, knock up, cramping is always advisable, using well-made hardwood cramping blocks. In cleaning off the joints use a very sharp plane and plane inwards to avoid the breaking off of end grain corners.

With careful adjusting this tool has successfully cut joints 300cm (12in) wide. At this width the work could do with greater support than the normal face will give, particularly when two pieces are on the face together. For this purpose a face extension has been planned which can be quickly attached with four woodscrews when required.

It is a wise plan to keep specimens of the joints made with the different wobble washers. These will be useful in planning future jobs. A different saw or differing amounts of set on the saw teeth will of course affect the spacing. Sizes have been deliberately omitted as the requirements dictated by each sawbench vary so much.

170 The disc cutter

Discs are required from time to time in most workshops, for purposes ranging from wheels for toys to small table tops. Everyone does not necessarily own a lathe and even turners find it an advantage to start off with a reasonably true disc. This tool will both saw and then sand discs with considerable accuracy and with safety. It is suitable for most small sawbenches and the construction is simple and straightforward.

Hardwood, multi-ply or blockboard are equally suitable to make a large baseboard as long as, or even longer than the length of the sawbench. A width of about 300mm (12in) seems to cope with most requirements. Screw on a mild steel strip to suit the sawbench groove. Two screws only will make sure that the strip is not distorted. Fix the metal in such a way that pushing the job across the saw will cut off the smallest amount from the baseboard. A turned knob or handle of some form is now fitted. Near each end of the base a row of holes is bored at about 25mm (1in) intervals. All these holes are threaded for a suitable screw, say 6mm ($\frac{1}{4}$in). The amount of wear is negligible so these threads will last for years. Each hole is countersunk on the underside. A machine screw coming from underneath will hold the knob handle in whatever position is later chosen as most convenient.

A pivot arm, a little longer than the base, is prepared from hardwood to about 45 × 13mm (1$\frac{3}{4}$ × $\frac{1}{2}$in), sizes are not critical, and it has a 6mm ($\frac{1}{4}$in) hole at one end. A machine screw is brought up through one of the holes at the handle end and fitted with two washers and a wing nut. A similar screw is put through at the other end.

On to this is fitted the stop bracket. Mild steel of 20 × 3mm ($\frac{3}{4}$ × $\frac{1}{8}$in) made the one illustrated. One end is cranked up for about 35mm (1$\frac{1}{2}$in). The arm has a 6mm ($\frac{1}{4}$in) slot made in it. The length of the slot must be a little greater than the spacing between the holes in the base. The construction is now complete and the tool is set up and operated as follows. Decide on the size of the disc required, draw it with pencil compasses and cut out a rough square containing this. Pin through the pivot arm, somewhere near its middle, into the centre of the drawn circle. At the handle end, bring up the screw through a convenient hole and when it is tightened firmly, drop on a small packing block made from an offcut from the disc. Add a washer, the pivot arm, another washer and the wing nut.

Put in a sharp saw and fit the stop bracket on to a convenient hole. A stout rubber band to a screw in the end hole pulls the arm on to the stop bracket. With the pivot arm held firmly against the stop bracket, pass the tool across the saw. Adjust the stop bracket until the cut is just clear of the pencil mark. In this manner continue to cut off corners until the work is virtually round. The workpiece must be held very firmly against the base to prevent any rotation when sawing. Be satisfied with quite small cuts.

The saw can now be replaced by a sanding disc and the stop bracket advanced by just that small amount. Sand down to the line by both rotating the disc and also sliding the tool, in this way spreading the wear on the disc. If a number of identical discs are to be made, saw them all first and

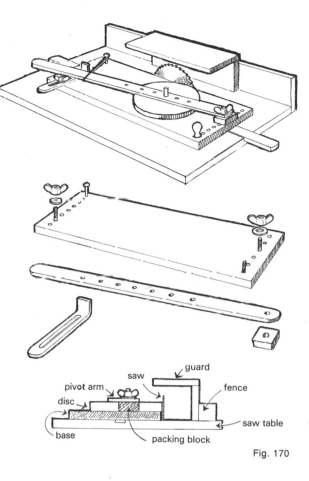

Fig. 170

then sand. Centre holes can be bored out on the pin pivot later.

The normal guard can probably be wedged up and arranged to suit the work in progress. If all else fails, a guard along the lines of the one illustrated will be quite satisfactory. As there is unavoidably a certain amount of leaning over the sawbench it is necessary for safety to remove loose clothing such as a necktie.

171 Shrinkage buttons on the sawbench

The buttons which hold on a table top are normally made by hand. Not only is this time consuming work but it assumes a supply of very wide, short-grain offcuts. Much timber these days of any real width, is warped. On the other hand workshops are littered with long thin offcuts, ideal material for making buttons. It was to utilise this material and mechanise the process that this device was designed. It is built mainly from blockboard and multi-ply. Sizes will have to be arranged to suit the sawbench on which the work will be done. The illustrations give broad details.

The baseboard **A** is fitted with the customary slide bar and is wide enough to overlap the saw by about 35mm (1½in). Two stop blocks will eventually be fitted underneath to restrict movement to the minimum necessary. In the early stages keep the baseboard much longer than the table. Fit it to the table and wind up the saw to cut the slot shown. The double fence is made up and fixed exactly at right-angles to the saw blade and touching it. The lower step should be 20mm (¾in) thick, no other dimensions are critical. A standard type of button for small coffee tables is shown in **B**. Before production can start, however, a double length-stop is re-

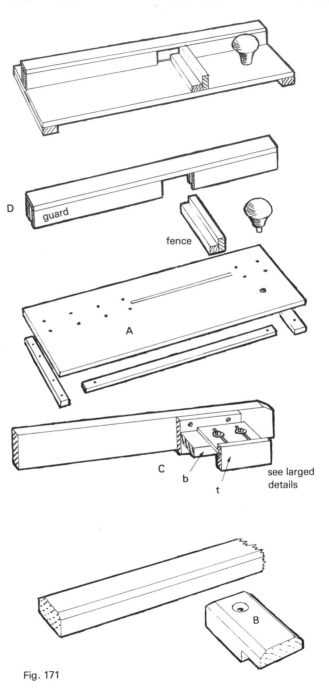

Fig. 171

quired giving length of tongue, **t**, and length of button, **b**. **C** shows this fixed to the normal wooden sawbench fence. A trial run can now be made.

Place a piece of prepared material on the step of the fence and adjust the height of the cross-cut saw blade until a tongue of the required thickness will be left. Now adjust the main sawbench fence until the length stop **b** will cut off buttons of the required overall length when the material is held on the baseboard. Clamp firmly and adjust the secondary length stop **t** to give the required length of tongue. The sequence of operations follows. With a strip of material on the baseboard, cut off a square end. Then place the strip up on the step and, pushed against stop **t**, make the first cut. Slightly withdraw the strip and make a second cut and repeat until the tongue is formed. With the strip on the baseboard, push against stop **b** and cut off the button.

The two stops below the baseboard can now be fixed to give sufficient movement and excess length sawn off the baseboard. This can be quite a dangerous operation so a guard is required as shown at **D**. The saw is thus totally guarded. The work is fed through quite a small gap and the completed buttons are poked away with a stick. A turned knob can now be fitted.

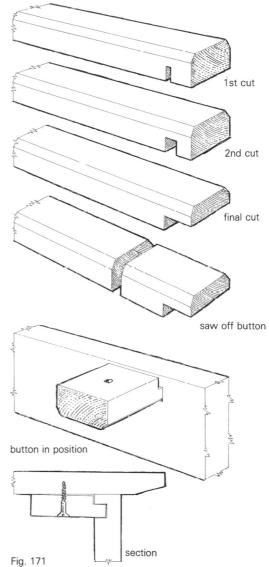

1st cut

2nd cut

final cut

saw off button

button in position

Fig. 171   section

It is convenient if material is first prepared with a small chamfer or rounding in order to cut up later into buttons shaped as in **B**. There are many further uses for the tool including multiple cutting of identical small components.

172 A grooving device for the sawbench

Two good hardwood blocks are the basic requirement for this tool, **A**. The bottom block is 20mm (¾in) thick and has a groove worked in it of 10mm (⅜in) width. This should be deep enough to take two thin brass or steel strips and two round-head screws. Before gluing on the narrower top block (which should be about 25mm (1in) thick) cut out a housing about 6mm (¼in) deep to take the renewable thrust block as shown. This prevents a ragged edge to the cut on the workpiece.

To this assembly secure a steel bar which fits the sawbench groove. One screw is countersunk but the other, a round-head with washer operates in a slot or over-large hole. This permits accurate adjustment of squareness between the tool and the saw. When secure take a pass over the saw, cutting not more than 6mm (¼in) into the upper block. The two strips which together form the locating teeth are out of thin brass or steel. Secure with two round-headed screws and washers.

To use, first prepare the wood for the grooved members accurately to size and, in addition, a few practice pieces of the same size. Move the locating teeth well out

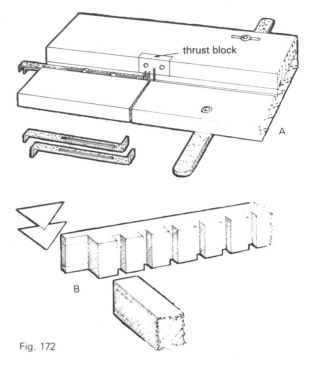

thrust block

Fig. 172

of the way and make the first cut in the required position. Move to the position for the second cut, with the teeth loosely located in the first cut. Hold the work firmly and push and pull the locating teeth to become a tight fit in the first cut. Screw them tightly in this position. Make the second cut then continue in this fashion, transferring each cut, when made, to the locating teeth.

Boxes can be made as shown in **B** by cutting the lap joint before grooving. Alternatively, loose grooved strips can be fitted to a box of any construction. When the depth of groove is the same as, or less than, that of the previous job, the thrust block should be changed in order to obtain a good finish to the groove. For grooves wider than the saw thickness, either use wobble washers or pack the saw on alternate sides with thin card to create the wobble, or drunken effect. A sharp combination saw is the ideal.

173 Photographic-slide boxes

The problem is that of making the grooved spacing blocks; 100 slides need 200 grooves and even if the reader were tenacious enough to cut these by hand and produce the finished box, he would certainly hesitate before undertaking another! The drawings show a jig which is easily made and will produce these grooved strips with ease and precision. Round these the rest of the job is built. The only essential is the use of a sawbench.

The jig consists of a bed and a slide, **A**. The bed has a groove cut down its centre to hold a strip of mild steel or brass 13 × 3mm ($\frac{1}{2}$ × $\frac{1}{8}$in) which projects no more than 3mm ($\frac{1}{8}$in) and may be glued in with a wood-metal glue, any sharp corners being filed off first. The saw slot is cut by the circular saw itself and should allow the saw to go a shade closer than 3mm ($\frac{1}{8}$in) to the metal strip. Various wood blocks can be glued underneath to position the bed exactly on the sawbench. If the

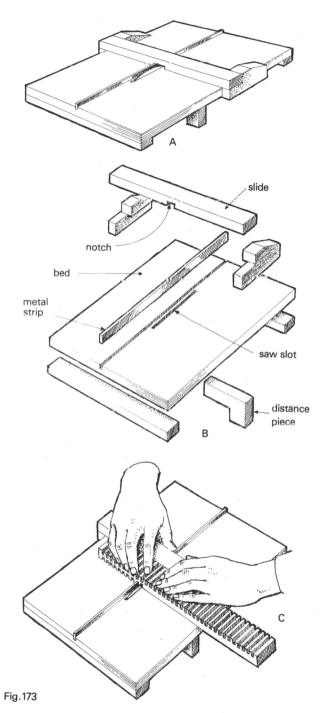

Fig. 173

179

saw table is large a distance piece, as shown, may be needed, as the bed itself need not be large. That illustrated is held firm by two G cramps. This problem of securing and later replacing in an identical position must be solved by each reader in relation to his own sawbench.

Details of the slide are obvious from the exploded drawing **B**. To reduce wear and friction the main cross member should just clear the bed and the slide should run either on the sawbench, if large enough, or on the rebates in the side pieces on a small sawbench.

There must be no side play and the slide should be as tight a fit on the bed as possible having regard to ease of working. For this reason a heavy plywood bed is better than solid wood as it is less liable to shrink. A notch is cut out large enough to clear both the saw and metal guide strip adequately. The slide is best assembled *in situ* by glue and screws. It can then be checked that the working face of the slide is square to the guide strip as this is a vital factor. The greater the length of the side pieces, the less is the tendency for the slide to get out of true with the guide. Having assembled the slide it can be removed to set and then be cleaned up.

The strip to be grooved must be planed exactly parallel and one end must be accurately squared. Most hardwoods will do, but beech or sycamore is to be preferred. A sharp cross-cut saw is needed and this must be packed up with two pieces of cardboard or with wobble washers on a very small thin saw until the groove produced is a tight sliding fit on the guide strip. This method can be used to set the saw guide on to the bed. Now set the height of the saw to give the depth of groove required. This is generally about 3mm ($\frac{1}{8}$in) but it must be slightly greater than the height of the guide strip to give clearance. Next the whole bed is slid on the sawbench until the distance between the sawcut and the guide is equal to the space required between grooves. The machine is now ready for production.

Place the square end of the strip up to the guide and against the slide, which then pushes forward over the saw. The slide comes back, but not the wood. The wood is now re-positioned against the slide but with the first groove on the guide. This is repeated as is clearly shown by **C**. When cutting the second slide care must be taken to begin at the true end of the strip. If desired, wide strips can be cut to be later ripped down into the required widths. Double-sided strips can conveniently be 30 × 20mm ($1\frac{1}{4}$ × $\frac{3}{4}$in) and single ones 30 × 13mm ($1\frac{1}{4}$ × $\frac{1}{2}$in). A shaving will remove any rag from the saw and the strips are cut, as a rule, to include 25 grooves, the blocks underneath the bed and the distance pieces ensuring that future stock is very close to the original. Lubricate as necessary with french chalk or paraffin wax. Wear on the slide can be taken up by a slip of veneer, planing the base again to give a working fit.

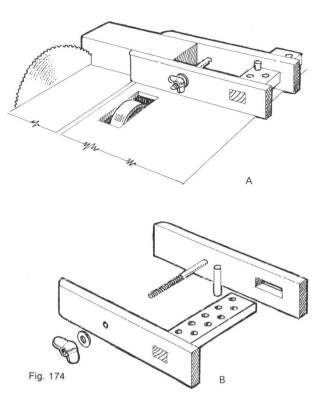

Fig. 174

174 Vice for the sawbench

The value of a circular saw lies in its precision as well as its speed. When properly used it will cut the wood square or at any required angle. It often happens, however, that a cut is required in a piece too small to hold with safety. The danger of handling small pieces, particularly for cross-cutting, is so obvious that few would attempt it and it is to overcome this problem and to eliminate the danger that this appliance has been devised.

In addition to square cutting it will do angle cutting up to about 10°. Beyond that the length of jaws would have to be reduced thereby rendering the vice less efficient for its purpose of square cutting. The vice is shown at **A** holding a small block and set to cut at 5° and **B** shows the vice dismantled. Sizes obviously cannot be given since they vary so much from bench to bench. The two jaws might conveniently be cut from 75 × 20mm (3 × ¾in)

stuff and the spacing bar from any suitable offcut.

The spacing bar is through-tenoned glued and wedged to the front jaw. The rear jaw has a mortise to accommodate this bar with a working fit. It is made slightly longer than the width of the spacing bar to give some necessary play in this direction. The spacing bar has 10mm (⅜in) holes drilled as close as possible without weakening the wood. A long 10mm (⅜in) coachbolt provides the screw. It is given extra length of thread and its head is well sunk. It is fitted with a wing nut and washer. A 10mm (⅜in) dowel peg complets the job, this being inserted in whichever hole suits the wood to be sawn. In practice use the nearest hole which will keep the jaws as parallel as possible. The vice should be held to the cross-slide of the sawbench with bolts and wing nuts or with wood screws. It should not be held loose.

A point to note is that if the jaws are not parallel the cut may not be square. This is because only the nearest jaw is actually fixed to the cross-slide. Ensure, therefore, that the wood is firmly in contact with this near jaw, inserting a little packing, should it be required. If there is any doubt, hold a try square between the near or fixed jaw and the sliding metal bar.

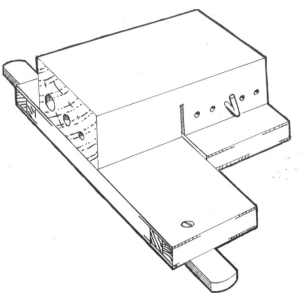

Fig. 175

175 Dowel cutter for the circular saw

A baseboard of 13mm (½in) plywood or 20mm (¾in) blockboard is fitted with a steel slide bar to fit the saw table. The board is passed over the saw to produce a true working edge and a stout hardwood block is glued to the base, square to the true edge. Three common-sized holes for dowels are drilled in the side of the block and the saw is set to the minimal height of cut which will cover each of the three sizes. From this saw cut a line of small holes is drilled to take the stop

pin. The same system is followed as described for the hand tool on page 96. The machine guard should be suitably adjusted or else a special guard should be made similar to that in Fig. 169.

176 Device for accurate angle and dimension cutting on the sawbench

Small components for mirrors, frames, etc. are often required in quantities for which hand sawing and mitring is too slow. This device will produce any number

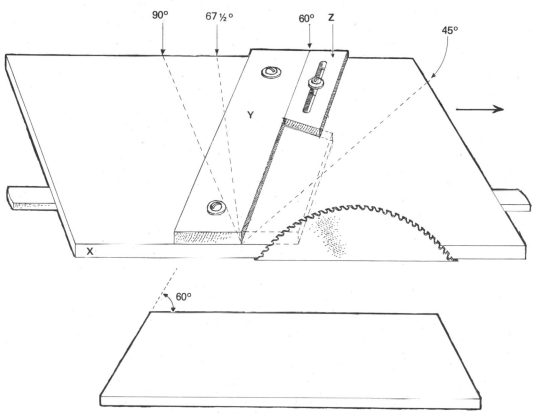

Fig. 176

Typical component considered

of accurate and identical components. A good blockboard or multi-ply base has a metal strip screwed to it to suit the sawbench groove. Now pass across the circular saw to produce an accurate working edge **X**.

The angle block **Y** is screwed through oversized holes, which permit slight adjustment to the base. This can be very accurately fixed for the common angles of 90°, 60° and 45° with a try square, or try square with set square. The angle of $67\frac{1}{2}°$, for an octagon, or

any other angle needs to be very accurately drawn, from the working edge. A length stop, **Z**, completes the tool.

Use a sharp, fine cross-cut saw blade and cut the components overlong. Now set the length stop so as to accurately cut one end of all the components. Reset the length stop and cut the second angle on all the components. Accurate setting will produce frames of four, six or eight sides which will glue up straight from the saw. See also page 103.

183

Fig. 177

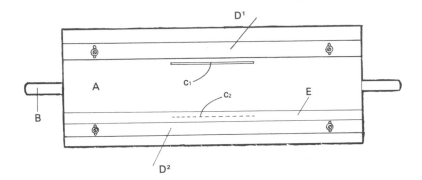

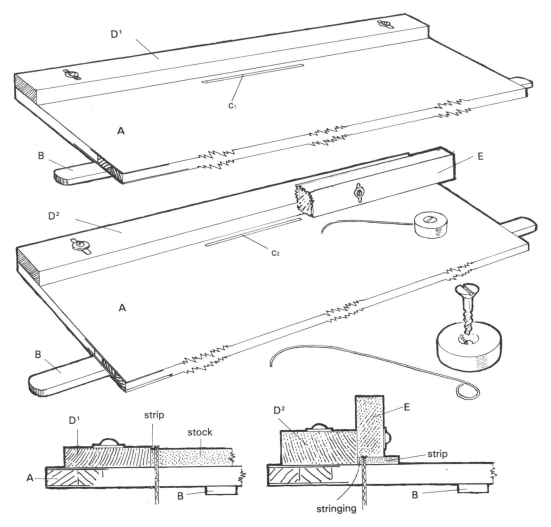

177 Inlay stringings on the sawbench

Inlay stringings can be bought but the varieties of timbers and sizes offered are very restricted. With this device they can be quickly and easily produced on the sawbench. A sharp and thin blade is required, preferably with the combination type of teeth. The device has been made double sided, as seen in the plan, so that rectangular and square section stringings can be made without resetting.

The base, **A**, is made from blockboard, as long as the sawbench and of suitable width. A steel bar, **B**, to fit the sawbench groove is screwed centrally below and this subsequently allows accurate replacement. The base is held in place with two G cramps and the saw is slowly wound up to cut through at C^1. Make this slot, C^1, as short as possible. The plain fence, D^1, is set up to give the thickness of the stringing. The illustration below left shows this in section. Saw off sufficient strips of this section.

Now turn the device round and cramp to bring fence D^2 into play. Make the saw slot C^2 as before and adjust fence D^2 to the width required. (This may or may not be the same as the thickness as D^1.) Adjust the holding down member **E** so that the work slides easily through. The saw must just

cut into this as shown in the section illustrated on the right.

The finished stringings are now sawn out. The workpiece can be held against the fence, either by a piano wire spring as shown or by a thin wood spring as described on page 146.

178 A mitre tool for the circular sawbench

Small boxes, particularly those to be veneered later, are often made with mitred corners. These are sometimes strengthened with veneer keys as described on page 25 and this tool enables one to cut the mitres both speedily and accurately.

The base is of thick multi-ply or blockboard to which is screwed a mild steel bar to fit the groove in the sawbench. Sizes must suit the particular sawbench and the size of work envisaged. At this stage the saw slot is cut.

The worktable is now secured to the baseboard by two glued 45° blocks. The angle can be accurately arranged by using a large drawing set square. The lower edge remains square and is lined up to the saw slot. This is cut vertical when the glue is dry by passing across the saw. A slim hardwood stop strip is fixed to the worktable. This is lined up with an accurate try square to be exactly at right-angles to the trimmed lower end. A home-made or purchased handle is fitted and the work stop completes the tool. This latter is a piece of cylindrical material, bored with a 6mm (¼in) hole well off centre, secured by a screw from beneath, with washer and wing nut.

To use this jig to make a box, first prepare the four sides accurately to size and provide one or two additional test pieces of the same thickness. Place a test piece in the work table and adjust the work stop until the cut is exactly on the corner. Then go into production. For greater accuracy cramp the piece to the work table.

The joint can be further strengthened by a loose tongue in two grooves. To achieve this remarkably strong little joint, after the mitre cuts have been made in some of the practice pieces, turn one over, adjusting the work stop and saw height to give a groove of the required depth in the right place. Depending on the width of the saw cut, an adjustment of the work stop and a second cut may be necessary. The tongue may be of thin ply or be a thin hard-wood strip. For appearance, if the box is later sawn through to give a lid, and for maximum strength, a cross-grained tongue is the best.

A device for planing thin strips accurately is given on page 80.

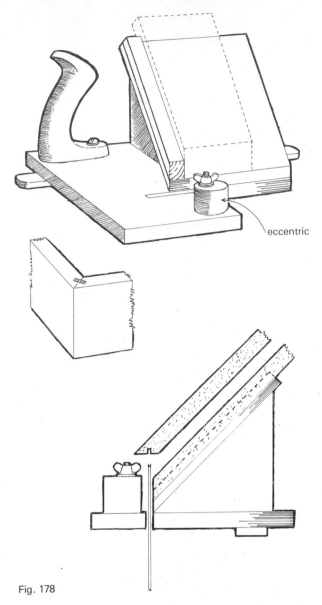

eccentric

Fig. 178

179 Sawbench tool to put keys into mitred frame joints

Picture frames are frequently strengthened by saw cuts into which veneer strips are glued as described on page 24. Larger frames with mitred corners can be similarly strengthened with stronger keys either of plywood or prepared hardwood strips. The saw cuts can easily be made on the circular sawbench.

The frame is mitre-jointed and glued up and at this stage the joint is quite fragile. The device the joint is held in for sawing is a false face to be screwed on to the face of the tenon tool described on page 153. Sizes will naturally have to fit the existing tenon tool. The false face can be from 13mm ($\frac{1}{2}$in) plywood, and the two fences can be of any convenient hardwood, say 20mm ($\frac{3}{4}$in). These are glued and pinned on at angles of 45° but it is more important that the angle between the fences is an exact 90°, which can be set up with a set square. The lower ends of the fences are liable to be cut during the operation, so pins or screws should not be put in too low. The workpiece is held between the fences and secured with bar, bolt and wing nut.

Choose a circular saw of convenient gauge and adjust the saw height to give the greatest depth

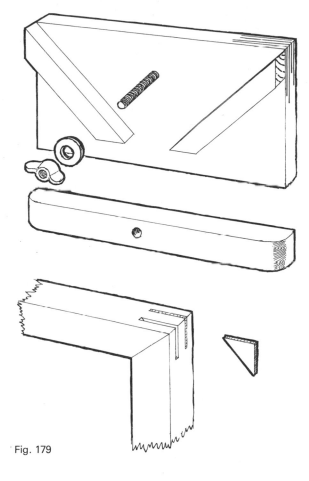

Fig. 179

of cut without coming through. Finally the tenon tool is adjusted to make the saw cut in the desired position. If required either make a second cut or adjust the fence to widen the cut to suit the material to be used for the key. Make the keys slightly oversize and clean up after gluing.

187

MISCELLANY

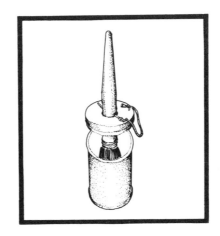

180 Sketching set square

The most common of the quick and easily drawn pictorial views is the 30° isometric shown at **B**. It suffers from two disadvantages. Often the unimportant side is given as much prominence as the interesting side. 45° mitres have a habit of disappearing or becoming confusing. A sketching set square based on the German DIN system is shown at **A**. It produces a much improved view as in **C**. It should be remembered that distances along the 42° axis should be reduced 1:0.66. These set squares can easily be made from $1\frac{1}{2}$mm ($\frac{1}{16}$in) clear polythene sheet. Bore a hole in the centre to make handling easier.

181 Dovetail set squares

A dovetail drawn in isometric projection is shown at **A**, (30° and 30°). A dovetail drawn to the DIN system described and illustrated in Fig. 180 (42° and 7°) is shown at **B**. The drawing of dovetails is a messy, time-consuming occupation involving one set square sliding on another, always with the liability of slip and error. These set squares eliminate all error and are speedy in use. They can be cut close to size then finished by hand planing.

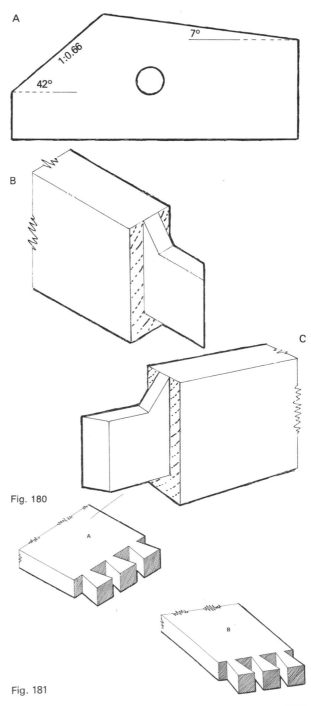

Fig. 180

Fig. 181

The isometric version, **C**, has a base of about 150mm (6in). The angles can be taken as $22\frac{1}{2}°$ and 39°. A short 90° at each end saves reaching for another set square for drawing verticals. A slope of 1 in 6 has been chosen as this shows up best in a pictorial view. With the DIN version, **D**, the angles can be taken as 31° and 56°. One right-angle has been included. A slope of 1 in 6 again shows up best. Allowance has been taken of the scale reduction of 0.6 along the 42° axis. A set square for orthographic projections, **E**, is another time-saver. A 1:6 slope has been chosen, which is approximately $9\frac{1}{2}°$, but 1:8 can easily be made.

A hole in the centre makes all these set squares easier to slide on the paper. The cut out in a conventional set square can filed away to make a dovetail set square as shown at **F**.

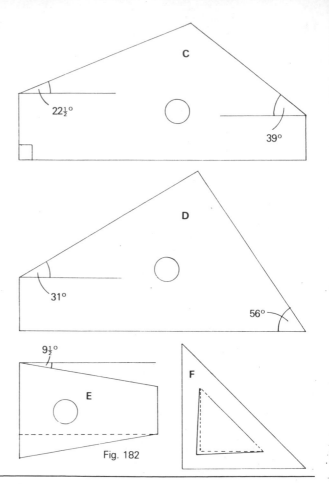

Fig. 182

182 A perspective drawing board

A perspective drawing is the ideal way to visualise a design or to show it to a client. Unfortunately a great deal of work is needed to produce one. This board will produce a perspective drawing to a scale and size suitable for drawing furniture.

First set out the whole scheme on decorators lining paper or on a roll of graph paper. The points may be constructed as in **A** or calculated by trigonometry. The following data must be fixed.

Distance of observer from object = O-CV
Height of eye = EL-GL
Scale ($\frac{1}{8}$ scale to imperial measure is most suitable)
Angle of object to picture plane (commonly 60° and 30°)

190

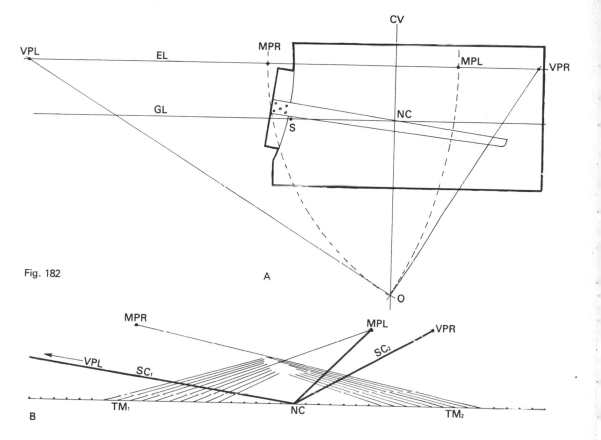

Fig. 182

A

B

Draw the eye line EL
Draw the ground line GL (to scale from EL – generally 5ft)
From O draw O-VPL (30°) and O-VPR (60°)
This fixes the left and right vanishing points.
With centre VPR and radius to O draw the arc giving MPR.
Repeat on VPL giving MPL.

These are the measuring points. Mark a suitable scale, **B**, on the ground line left and right of CV (centre of vision), making zero at NC (nearest corner).

Join NC to the two vanishing points. These are the ground lines of the drawing. Fix strips of card along these lines, then join the graduations of the left-hand side to MPL. This gives the diminishing scale SC^1 on the card. Repeat joining the right-hand graduations to MPR, giving the diminishing scale SC^2 on the second card.

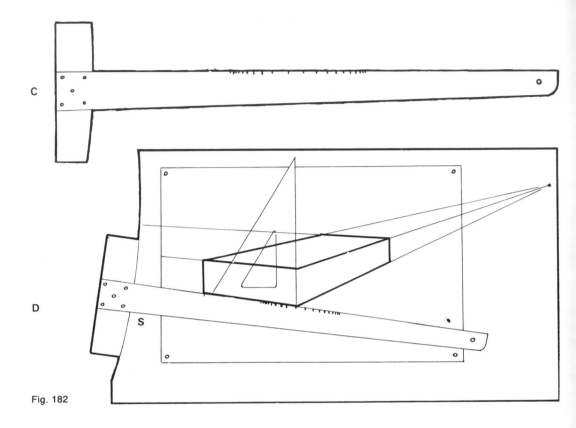

C

D

S

Fig. 182

These diminishing scales can later be made more permanent or scratched on to the T square, **C**. Now draw in the shape of the board, transfer to wood and cut out. Make a T square to match, **D**. Mark permanently on the board CV-O and NC; NC-VPR; NC-VPL and drive in a pin for VPR.

Fix a stop pin or block for S ensuring that this lines up the T square on NC. A 60/30 set square must be cut away so that

with the T square on its stop, lined up on VPL-NC the set square lines up on CV-O. All verticals are drawn with this set square while the T square is on the stop. The scales can be transferred to the T square edge in such a way that it can be used as a rule to VPR without the stock being in the way. All measurements must be made only on NC-VPR, NC-VPL, NC-CV and projected to where required. Further operation is obvious.

192

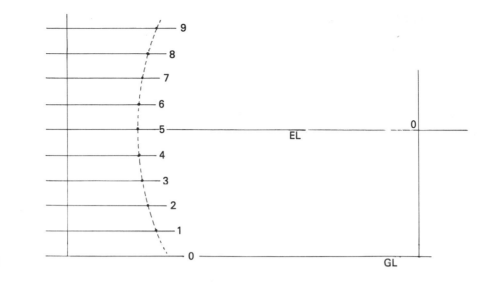

Fig. 182

Refinement of vertical scale, **E**.
Draw O-EL as O.NC on **A**.
Draw in GL as before GL-O = eye height.
Draw arc O-EL.

If eye height is (5ft), divide the lower arc into 5 and repeat above the eye line.

Project horizontally to give a scale diminishing above and below eye level.

The vertical scale can be neglected and quite a reasonable drawing still produced.

Popular Formulae (imperial sizes)

	A	B
Distance	12'	3'
Eye Height	5'	$1\frac{1}{2}$'
Angle	60°	60°
Scale	$\frac{1}{8}$	full

Metric scales $\frac{1}{5}$ and $\frac{1}{10}$ are respectively rather too large and rather too small for furniture.

A true or diminishing vertical scale can be scratched on the underside of the set square. Giant perspective details are drawn by using a formula like **B** with a scale of say X8. The V.Ps would then be calculated by trigonometry and long chalk lines used.

183 Small blocks by disc sanding

At times small blocks are required which are too small to plane and yet need a better finish than off the saw. The false table may be of blockboard and a stop block is fitted so that the board, when placed on the sanding table, will just clear the disc. Rough-sawn work has a side sanded true first, then a right-angle at one end. The latter can be done outside the length blocks. A block **a** is pinned in place to produce the required width. The work is placed here and the false table pushed forward until it is stopped. The length is produced similarly at **b**.

184 Forming circles on the disc sander

Accurate discs can easily be made without a lathe using either a disc sander, or a sanding disc in the circular saw. A basic false table is required, usually from thick multi-ply or blockboard. A hardwood fence of about 20mm ($\frac{3}{4}$in) is glued to this to give such an overlap on the table top that the end of the false table just clears the sanding disc. This fence is drilled and tapped to take two 6mm ($\frac{1}{4}$in) screws. Thumbscrews are most convenient but slotted heads will work equally well.

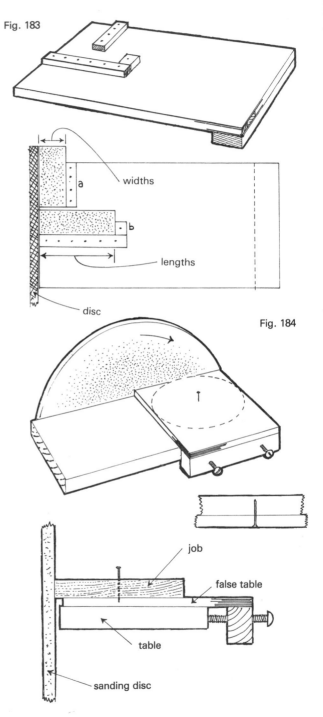

Fig. 183

widths

a

b

lengths

disc

Fig. 184

job

false table

table

sanding disc

To use, first cut out the discs as carefully as possible either by hand, bandsaw or jigsaw. One to two milimetres oversize is enough. The adjusting screws are set to protrude somewhat more than this allowance. Set the false table on to the machine with the screws hard against the table edge. With this held firm, pin on the disc through its centre with its circumference hard against the sanding disc. Pull back the false table and switch on. Push in the false table until it will go no further and slowly revolve the disc. In order not to burn the job and wear out the disc it is necessary to revolve the disc and slide the table at the same time. This ensures that fresh wood is meeting fresh glasspaper all the time. When no further cut is possible, withdraw both screws and repeat. Continue in this manner until down to size. If it is required to fit the disc to an opening, clip off the pinhead before starting. In this way the disc can easily be removed and replaced. If a centre hole is not acceptable the disc can be pinned halfway through from beneath.

185 Pattern sanding

This is a method by which the disc sander can make small components and repeat them accurately to size. A false table is required having a thin upstanding metal edge. This should be screwed to the sander table as

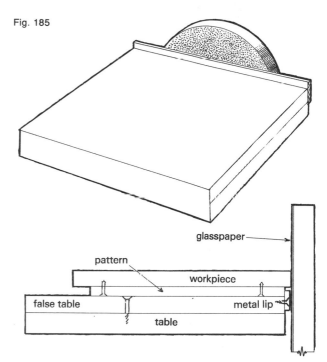

Fig. 185

glasspaper

pattern

workpiece

false table

metal lip

table

close as possible to the disc itself without fouling it. A pattern of the required component is made slightly undersized and the distance from the inside lip on the false table to the disc itself is, in fact, the amount that the pattern should be undersize. Hardboard, being of even thickness and easily worked on the edge, is a good material for the pattern.

The workpiece is cut slightly oversize and then is pinned on its underside to the pattern. The work is sanded to size, the sanding stopping as soon as the pattern reaches the metal tip. Any straight or convex shape can be made but, naturally, a concave shape cannot. See also Fig. 160L.

195

186 Boring paper

Pads of paper or booklets can be effectively and cleanly bored by clamping the paper firmly between two stout pieces of wood then boring through with a sharp shell bit.

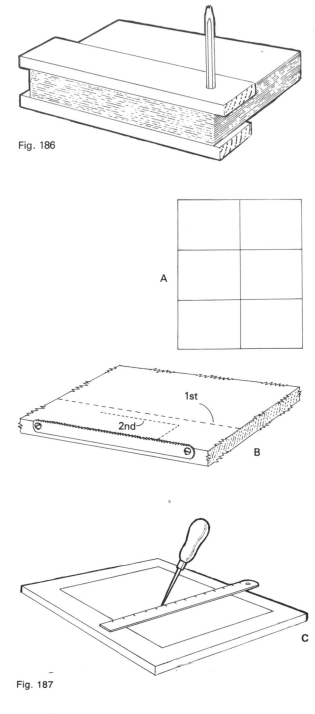

Fig. 186

187 Glasspaper tearing

Fortunately the standard sheet divides exactly into six pieces which fit the cork glasspaper blocks. These pieces can easily be torn, without waste, by using this simple device. A 300mm (12in) hacksaw blade is screwed to the edge of a suitable block and two lines show where the sheet is positioned for the tearing. Glasspaper can be cut without damage to the cutting implement by using a scriber or compass point and a straightedge. The paper is cut by a scissors effect between the scriber and glass crystals. This action sharpens the scriber itself.

188 Oilstone case

With modern synthetic resin glues it is no longer necessary to carve an oilstone case out of the solid. An improvement on the traditional form is shown at **A**. A small end grain block at each end permits the sharpener to use the full length of the stone thus preventing it from wearing hollow

Fig. 187

in the middle. This can be extended at one end to give a bearing surface for a honing tool which would otherwise only operate on about two thirds of stone.

After honing edge tools a slight burr or wire edge will sometimes remain. This can easily be removed by wiping through an end grain block as at **B**. This can be permanently screwed to the sharpening bench or held in the vice. It is quite dangerous to hold the block in the hand.

In a communal workshop or where a large number of plane cutters are to be sharpened, this block, **C**, screwed to the sharpening bench will be found to speed up the process of removing the cap iron.

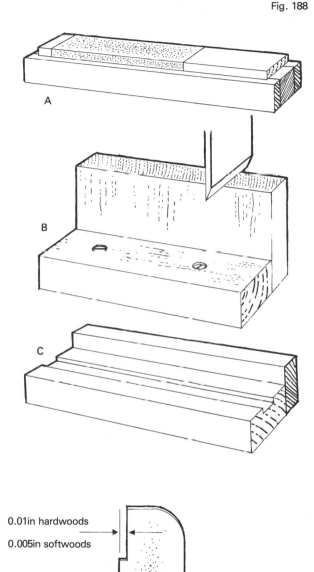

Fig. 188

189 Circular saw setting gauge for small saws

This gauge is best made from mild steel 2–3mm ($\frac{1}{16}$–$\frac{1}{8}$in) thick. The vital notch is filed out to a depth for hardwood saws of 0.250mm (0.01in) or for softwood saws of 0.125mm (0.005in). This is checked during filing with a feeler gauge on a truly flat surface such as a plane sole. A hanging hole should help to prevent loss. In use, bend over the teeth with the simplest lever type, set and check the amount with the gauge.

0.01in hardwoods

0.005in softwoods

Fig. 189

190 The bag press

Both veneering and laminating can be beautifully carried out using atmospheric pressure in a vacuum bag. The thick rubber bag, compressor/extractor and electric motor are all expensive items. However, in the smaller sizes, quite a good job can be done using a strong, preferably clear, polythene bag and a domestic or industrial vacuum cleaner. The domestic model should produce pressures of 12 pounds per square inch, the industrial will be somewhat better.

A section through the assembly is shown at **A**. The bag of, say, 1.0m (3ft) or 1.3m (4ft) is sealed at the mouth by a pair of cramped battens. To extract the air a plastic plumbing fitting is attached of the type which would connect to a thin sheet metal tank, not a thick ceramic fitting. Two thick rubber washers are fitted to make an airtight seal and a suitable rubber pipe connects to the vacuum cleaner fitting. This pipe can be sealed with two wood blocks and a G cramp. Alternatively metal pipes can be used with a normal water stopcock fitted. To keep the vacuum this must be one with a rubber tap washer.

A baseboard should be made from 20mm (¾in) blockboard or chipboard. The surface should be covered with 100mm (4in) squares made from hardboard with a gap between them of about 15mm (½in). At the intersections of these gaps a 5mm (¼in) hole should be drilled.

The job to be veneered should be inside a fold of thin clear polythene to keep the main bag clear of glue. Shaped work can be held together with adhesive tape, rubber bands or even by hand from the outside, the clear bag enables any movement of misalignment to be seen. Sharp corners must be avoided and triangular fillets tacked to the baseboard where there are sharp internal angles. P.V.A. glues tend to thicken before the full pressure is achieved. Synthetic resin glues are quite satisfactory, as are cold-setting scotch glues.

191 Cutting plastic and rubber foam

Foam upholstered, drop-in seats usually have rounded, cushioned edges. These require the foam to be cut at 45° not an easy task attempted freehand. The device ensures an accurate 45° with no particular skill being required. A plywood or hardboard offcut is required about 200–300mm (10–12in) wide and a bit longer than the foam. To this is glued a 45° strip, **A**, cut from approximately 50 × 50mm (2 × 2in). The foam can be firmly held by hand while a sharp plane iron, running across the angled strip, will accurately

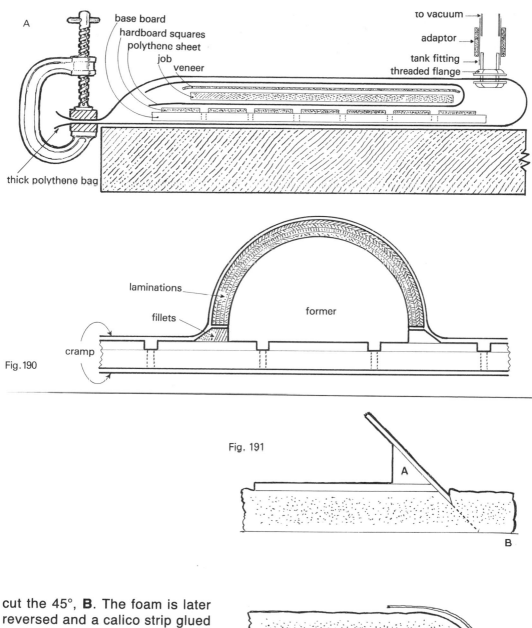

A

base board
hardboard squares
polythene sheet
job
veneer

to vacuum
adaptor
tank fitting
threaded flange

thick polythene bag

laminations

fillets

former

cramp

Fig. 190

Fig. 191

A

B

cut the 45°, **B**. The foam is later reversed and a calico strip glued to it with an impact glue. This is then pulled tightly round the corner of the base where it is tacked or stapled, **C**.

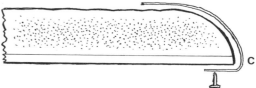

C

199

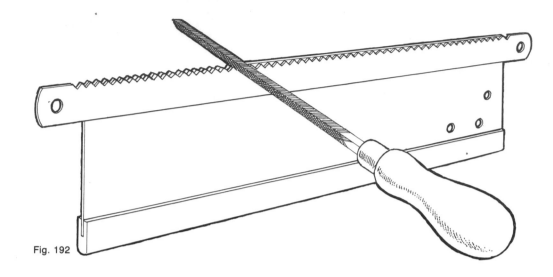

Fig. 192

192 Recutting saws

The rip tenon saw described in Fig. 131 is not commercially available so an existing saw must be recut at 10 points per inch (or 25mm). Nowadays commercial saw sharpening is so badly done that perfectly good saws must often be recut. Many firms offer to sharpen but few will undertake to recut. Fortunately there is an easy way round what appears to be a difficult and highly-skilled operation. All that is required is a saw vice and a high-speed steel power hacksaw blade.

File off all the teeth producing a smooth straight edge and then put the toothless blade and the hacksaw blade in the saw vice. Lock the vice with them both at the same height. With a new file, file straight across at 90° until the file is stopped by the hacksaw blade. It is a help to first colour the hacksaw teeth with engineer's blue, stopping as soon as the blue is removed. The normal setting and sharpening process follows from here. Not much will be left of the file after contact with the high-speed steel but this is a cheap price to pay for a recut. Power hacksaws are available in 10, 12 and smaller teeth per inch. There are now good saw filing guides on the market so no-one need fear the remaining processes. Similarly there are good pliers-type saw sets but beware of buying one which will not cope with teeth smaller than 12 teeth per inch (25mm).

If the angle of the teeth given by the power hacksaw is not exactly what is required the final sharpening with a saw filing guide will correct this.

193 Painting sticks

The difficulty of painting or varnishing woodwork is that frequently all the surfaces need coating, often with three coats. If one surface is left uncoated, for the job to stand on, the number of applications is doubled and a joint line is created between the two applications. The painting sticks avoid this. Two pieces of hardwood are chosen and into each are drilled four holes, two to take the spikes and two as storage holes for spikes when the sticks are not in use, **A**. To avoid rust the spikes should be of stainless steel, brass or alloy and they should not be aggressively sharp.

In varnishing, say, a table top which must be covered on both sides to prevent warping, the routine is as follows. Stand the top face down on two wood blocks, not the painting sticks and varnish the edges. Next cover what will be the underside. Now turn over and stand the painted surface lightly on the spikes and paint the top surface. The spikes will leave four tiny blemishes which will sand off. On the second and third coats these blemishes should be in different places.

After use, store with the spikes protected. One piece should have a centre hole with a cord, which has been secured with a

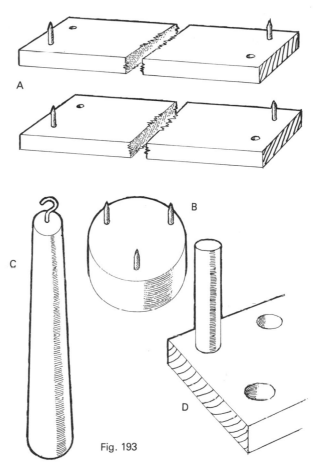

Fig. 193

dab of glue. Use this to tie the two sticks together.

Small components and pieces of turnery are best handled on a single block having three spikes, as at **B**. Using three spikes, the work will not rock during the painting.

Other items such as table legs can be hung from small hooks **C** or stood in holes drilled in a stout block **D**.

194 Wet storage of brushes

Much time is wasted by elaborate cleaning and drying of brushes which are in frequent use; for example, the brushes used for shellac sealer or clear polyurethane varnish in a communal workshop. Use a standard-sized metal tin or something similar, turn a lid to fit and bore a hole for the brush. A clip made from piano wire or spring steel holds the brush handle at its narrowest point. Alternatively drill the handle and suspend the brush from a strong wire pin. Tie the pin or the clip to the lid to prevent loss. Keep sufficient sealer, or in the case of the varnish brush, white spirit, in the tin to just cover the bristles. After use, squeeze out surplus varnish on to waste paper and store suspended in the white spirit. Change the white spirit from time to time.

195 Shouldered dowels

Small racks of various kinds are often built using dowels for rails, as in **A**. The lack of shoulders, **a**, makes accurate gluing up difficult. A rail with a shoulder and tenon, **b**, is a much more satisfactory job and this can be arranged by using a loose tenon, **c**.

A simple device for drilling the dowel ends to take the tenons is shown at **B**. It can be turned as

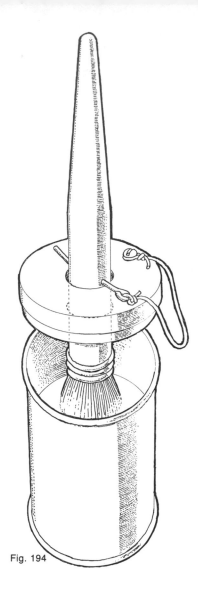

Fig. 194

202

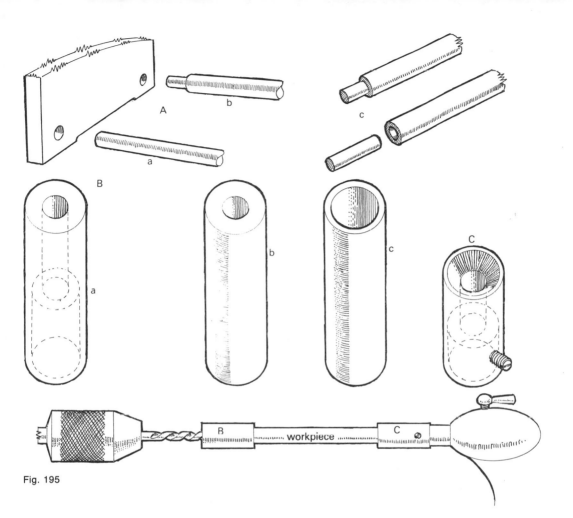

Fig. 195

in, **a**, either from wood, or better still, from mild steel. About 60mm ($2\frac{1}{2}$in) is long enough to accurately guide the drill. One end is drilled to suit the tenon, **b**, the other end, **c**, is drilled to suit the dowel. If much use is foreseen, the drill bore may be case hardened, see page 204. An electric drill is ideal for the job with the dowel held in the vice and the jig held still with the other hand.

If preferred, the work can be done in the lathe. This will require a concave tailstock fitting, **C**. If this is made from 25mm (1in) diameter metal, it will cope with the common sizes. An angle of about 45° is suitable. The other end is bored to take the tailstock and fitted with a small brass locking screw. When held in **C** the dowel is unlikely to turn during the drilling.

196 Case hardening

Some of the devices described in this book need a hard steel finish. This can be achieved by case hardening. Only mild steel can be case hardened. When hot the mild steel absorbs carbon, producing a skin of carbon rich steel over a core of soft steel.

Heat the area to a cherry red then dip it in case hardening component and leave for one minute. In the case of a bore, fill with compound and wait a minute. Shake off excess powder and repeat these two processes twice. Reheat to cherry red then quench in cold water. Do not grind or file the hardened area. Use nothing more than emery cloth. A good tool merchant will stock case hardening compound.

Hardening and tempering Mild steel will not harden and temper – it will only case harden. Hardening is only possible with tool steel such as is found in the cutting tools, scraper blades and saws. It can be bought in small quantities in a range of sizes from a good tool dealer and it is sometimes sold under the name of gauge plate.

File or grind the metal to the required size and shape and heat to a bright cherry red. Quench in either water or oil, moving continuously until all activity in the liquid ceases. The tool is now glass hard, generally too hard for use, and quite brittle and it needs to be tempered. Brighten with emery cloth the area which is required to be hard. Gently warm and look for the colour change. For the majority of edge tools, scraper and scratch tools a dark straw is the colour required. Scribers can be taken to yellow, screwdrivers and punches to purple and drills to brown. To heat small items evenly, lay them on a hot firebrick or a larger piece of steel. Watch very carefully, then when the desired colour is reached, quench in oil or water. After tempering, oilstone sharpening and grinding are both possible.

197 Steaming small components

Comparatively little steaming is undertaken possibly because of the rather elaborate and inefficient apparatus generally used to generate the steam and to contain the work. The simple apparatus shown here obviates most of the difficulties. A 150–200mm (6–8in) galvanised tube of about 1.800m (5ft) with a bottom soldered to it is required. A hole is cut to take a conventional domestic kettle element. The lid is a metal disc containing a hook from which to hang the work. With this lid in place steam condenses on it, and at the top of the tube, thus generally no topping up is necessary and surprisingly little water is required. Work is hung clear of the water by a string from the lid hook.

198 Carpet blocks

Polished surfaces need constant care and protection while in the workshop. To prevent damage from the bench top they should be stood on wood blocks covered with carpet offcuts. These should always be made in pairs and several assorted sizes will be useful. Use conventional carpet rather than rubber or foam backed and cut oversize. Almost any glue will do and when it is dry, trim the carpet to size with a sharp tool.

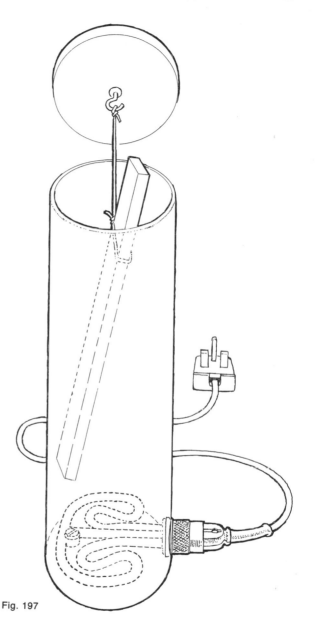

Fig. 197

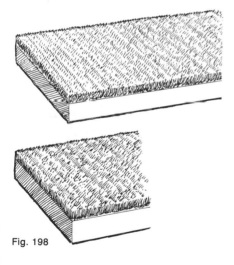

Fig. 198

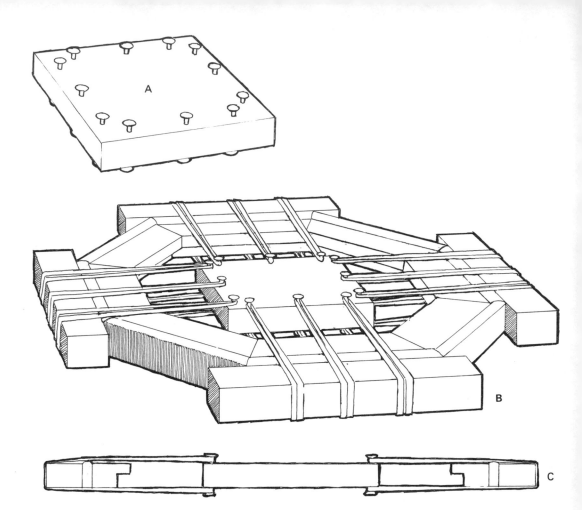

Fig. 199

199 Lipping frames

Light frames, square, rectangular, hexagonal or octagonal, often require a lipping, in some cases to conceal a loose tongue. They are usually too light in construction to support the weight of a number of sash cramps. This device is very light in weight and permits several lippings to be applied at one time, usually alternate pieces. Hardwood blocks are made slightly longer than the lipping and of about the same thickness as the lippings width. These are faced with an adhesive tape to prevent sticking after gluing. Width should be greater than this thickness.

A centre block is prepared as in **A**. For a square, rectangular and hexagonal frame a four-sided block is needed. The hexagon requires a hexagonal or triangular block. Tacks round the edges take a series of lacing rubber bands. **B** shows this lacing in plan, while **C** shows a centre section. Gluing alternate pieces enables the intervening pieces to be fitted later with precision.

200 Transparent adhesive tape

A Cramp blocks faced or covered with tape will not stick to a glued-up job.
B Tape in the corners of box constructions makes removal of excess glue easy.

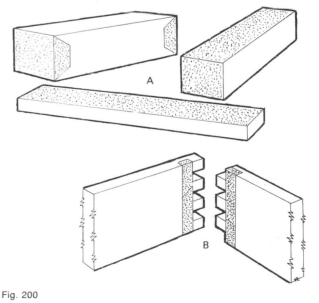

Fig. 200

C Joints such as dovetails can be protected when wax or oil polishing by a strip of tape. Any grease on the joint would impair gluing efficiency.

D Tenons can similarly be protected by wrapping in tape.

E A mortise can be protected from wax by a tape strip. Stick on a piece over wide then cut off to the exact rail size with a cutting gauge. This makes sure there is wax right into the corner, preventing excess glue from sticking.

F Thin lines can easily be applied by painting between two tape strips.

G Tenons on turned legs can be protected when polishing by a tape ring.

H Excess glue is kept off turned legs by a band of tape.

J Veneers and plastic laminates sometimes splinter when drilled. This can be reduced or eliminated by drilling through a patch of tape.

K Small splits can be held together with tape after gluing with a quick-setting glue. If necessary, a small wood block can be applied. This will not adhere to the job.

L Semi-permanent labels can be made by covering written or typed work with a wide tape strip. This is useful in a communal workshop for instructions.

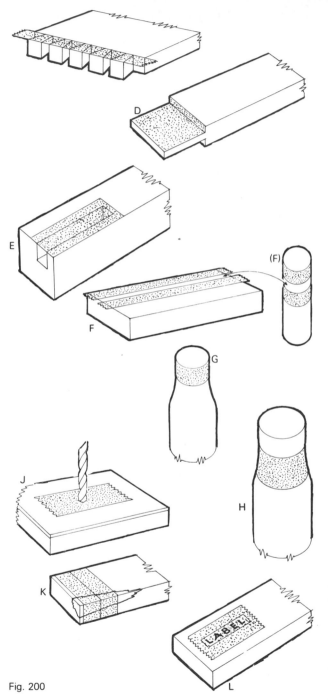

Fig. 200

208

201 Toothing plane cutter

Since the publication of this book, the supply of new and second hand cutters suitable for the toothing plane described on page 93 has virtually ceased. As the demand for the plane increases, for veneering and the laying of plastic laminates, I was forced to search for alternatives. As many workers tooth with hand-held pieces of hacksaw blades, this seemed to be the avenue to explore. The result shown in Fig. 201 has proved to be perfectly satisfactory.

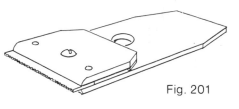

Fig. 201

The main plate, **A**. is from bright drawn mild steel. It has a large hole to accommodate the large woodscrew which secures the wedge. It is drilled and tapped to take the clamping screw, **B**, and two short locating pins, **C**. The clamping plate, **D**, is made from the same material. There are clearance holes for the clamping screw for the two locating pins. The lower end is rounded in the manner of a cap-iron. At the top end there is a spacing strip of about the same thickness as the hacksaw blade. This may be soft soldered or fitted by two small rivets. It is subject to no strain.

A wedge, **E**, is needed, rather similar to that of the Stanley type iron planes. The latter can be used at a pinch, but preferably it should be somewhat shorter. The clamping screw, **F**, is an improvement on the lever because its grip is more secure and positive. See page 210 for details of its construction. A non-ferrous foundry will cast the capiron in brass or aluminium from a simple wood pattern. Failing this, a wedge can be made from dense hardwood with a nut let in for the clamping screw.

The cutter, **G**, can be snapped off in the vice from a hacksaw blade. When these break naturally, there is often a relatively unused two inches at each end. Choose the coarseness, (the t.p.i. or number of teeth per inch) to suit the work in hand. Grind the ends to make the same width as the main plate.

It is not thought that an adjusting mechanism in a toothing plane is really worth while. However, if it is preferred, there is ample space in the body to accept one.

When the tool has been tried out and found to be satisfactory, finish all the parts to a high standard by

209

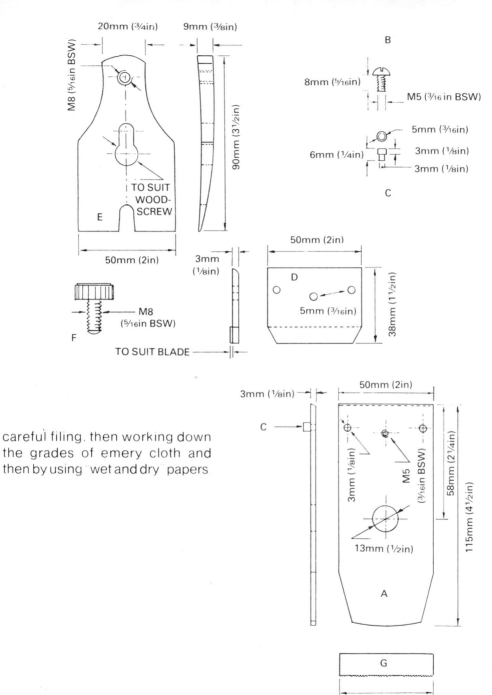

careful filing, then working down the grades of emery cloth and then by using "wet and dry" papers.

Fig. 201

202 Knobs

A number of the devices described here need knobs for clamping or for adjustment. From time to time knobs are missing or become damaged. It is not particularly difficult to make these. Two distinct forms are possible   to make in wood or in a synthetic material.

For wooden knobs choose a dense, hard wood such as ebony, rosewood &c. Choose by the suitability of individual pieces rather than by species. Knobs tend to be turned though they can of course be bench made. Cut out a block, oversize, drill and tap it centrally and then clean up the inside face. On the end of the metal screw, file two considerable flats, **A**. Thoroughly degrease. Coat the thread well with an epoxy resin adhesive and assemble, **B**. Pack any gaps well with adhesive, leaving it slightly proud. Give ample time for the adhesive to harden thoroughly.

The turning can be done in a wood or a metal lathe. If the former, keep the metal screw below the surface and turn down almost to it. In a metal lathe, finish by taking a cut across both wood and metal. When a number of knobs are to be made, a holder, as shown at **C**, will prevent damage to the thread by the chuck jaws. When turning the inside face of a knob using the holder, a lock nut or locking screw will prevent move-ment of the screw, **D**. Internally threaded knobs can of course be reversed. Using the power of the lathe, a durable finish can quickly be built up with linseed oil.

Synthetic knobs can be produced from a number of materials. Satisfactory ones have been made from synthetic resin glues such as Cascamite and Aerolite. Give ample time, i.e. several days, perhaps near a radiator, for the chemical action to become complete. Motorcar body filler is equally successful and much quicker but the most successful is fibreglass resin, which can be coloured. Care must be taken to avoid bubbles as far as possible. A successful mould can be made by boring a large hole in a wooden block, followed by a smaller tapped hole to suit the screw, **E**, which can be flattened at the end either by filing or by hammering, **F**. When the casting is well hardened, split the block to release the knob.

A more sophisticated and permanent mould can be made by using rigid plastic water pipe in which slides a well fitting turned wood block, threaded to take the screw, **G**. A number of blocks may be made to take different screws. The tube and block are oiled or waxed to permit easy release of the finished casting and the block can be adjusted to give knobs of different thicknesses.

These synthetics turn best at a

fairly high speed, preferably in a metalworking lathe, using a trailing cut. In a wood lathe, scrape, using a tool with a steep angle as for metal turning. Polish with successively finer grades of glasspaper and finally with "wet and dry" paper.

Internally threaded knobs can be produced in the same moulds, **G**. A nut, preferably a square one, is suitably positioned on the screw. In all cases where the mould is to be handled by a second person, it is advisable to fit a lock nut below the wood block, thus preventing accidental movement of the screw.

A large number of shapes are possible and readers will no doubt wish to create some of their own. A few are suggested at **H**. Flats can be accurately filed on turned knobs using the simple jig illustrated at **J**.

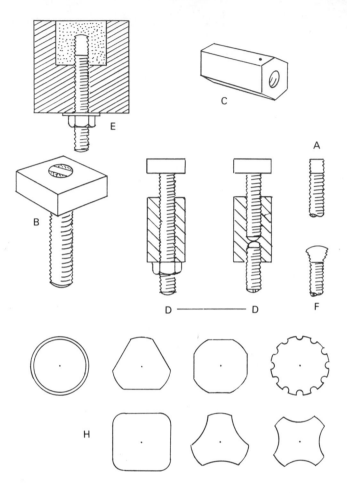

Fig. 202

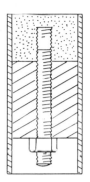
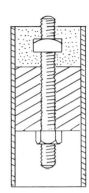

G   G

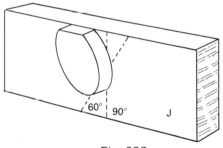

60°   90°   J

Fig. 202

203 Carcase cramping system

The high cost of sash cramps, particularly in the smaller sizes is one reason for making up this kit. A great reduction in the weight of ironmongery, especially in the case of smaller jobs, is another. Additionally the system is very handy for a craftsman working without help.

A ½in BSW thread has been chosen because it is the fastest thread commonly available but metric M12 is nearly as good. Being in tension this thickness is adequate. The following is a recommended kit.

8 pieces	304mm x M12 (12 x ½in BSW)	**A**
4 pieces	152mm x M12 (6 x ½in BSW)	**B**
8 washers	12mm (½in) i/d	**C**
8 nuts	M12 (½in BSW)	**D**
4 spinner nuts	to be made – see diagram	**E**
8 connector nuts	to be made – see diagram	**F**

supplemented by
4 pieces   914mm x M12
              (36in x ½in BSW)

Cramping blocks, **G**, from 50 x 38 mm (2 x 1½in) and 50 x 76 mm (2 x 3in) to suit individual projects.

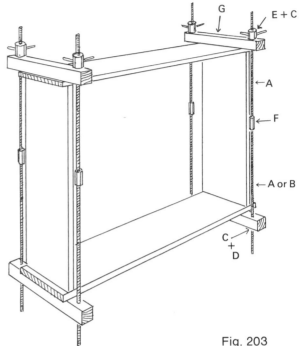

Fig. 203

The screw lengths and common nuts are straightforward. Connector nuts. **F**. must be made either from a hexagon or round section bar. The spinner nuts need plenty of weight. It is most convenient to drill the large hole and then drill and braze in the tommy bars, passing them right through. Finally, drill again to clear the bar and tap.

Connector nuts can sometimes be bought to order, but they are expensive. Readers in the USA are able to buy 100mm (4in) threaded connectors to suit their own threads in hardware shops. These can be sawn in two.

The blocks. **G**. are prepared with several holes, slightly oversize for ease of working and with one face slightly curved to give good pressure in the centre of a carcase if required. After a while a useful stock of these blocks mounts up.

Fig. 203 shows a basic carcase being cramped up. Note that if this stands on, say, two sawing horses, one man can cramp up the whole job unaided and without panic.

Once so equipped, the reader will rapidly find further uses for the kit.

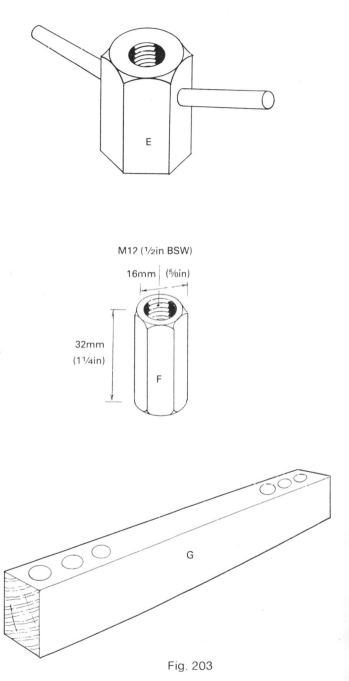

E

M12 (½in BSW)

16mm (⅝in)

32mm (1¼in)

F

G

Fig. 203

Index